CHILTON'S COMPLETE HOME WIRING & LIGHTING GUIDE

CHILTON'S COMPLETE HOME WIRING & LIGHTING GUIDE

L. DONALD MEYERS

CHILTON BOOK COMPANY
RADNOR, PENNSYLVANIA

The author extends his appreciation to the many individuals and companies who helped in the preparation of this book by supplying information and artwork. Special thanks are due to Kenny Mann for his technical assistance.

Published in Radnor, Pennsylvania, by Chilton Book Company
and simultaneously in Don Mills, Ontario, Canada,
by Nelson Canada Limited

Library of Congress Catalog Card No. 80-971
ISBN 0-8019-6790-2 *hardcover*
ISBN 0-8019-6791-0 *paperback*

Designed by William E. Lickfield
Manufactured in the United States of America

SAFETY NOTICE

Proper procedures are vital to the safe, reliable execution of all electrical installations and repairs, as well as to the personal safety of those performing repairs. This book gives instructions for safe, effective methods. The procedures contain warnings which must be followed, along with standard safety practices, to eliminate the possibility of personal injury or improper service which could compromise the safety of electrical work.

It is important to note that repair procedures and techniques, tools and materials, as well as the skill and experience of the individual performing the work, vary widely. It is not possible to anticipate all conceivable ways or conditions under which electrical work may be done, or to provide warnings against all possible hazards. Standard and accepted safety precautions and equipment should be observed, and safety goggles or other protection used during cutting, grinding, prying, or any other process that can cause material removal or projectiles.

Some procedures require the use of tools specially designed for a specific purpose. Before substituting another tool or procedure, you must be completely satisfied that neither your personal safety nor the quality of work will be endangered.

2 3 4 5 6 7 8 9 0 9 8 7 6 5 4 3 2 1

Contents

1

The Safe, Sensible Way to Do it Yourself

It is often said that life's pleasures are either dangerous, illegal, or fattening. Electricity is not ordinarily included in such pleasures. Love, sex, food, drink, a sunny day, and numerous other enjoyments come to mind first. But, if we think about it, electricity is a *source* of many of our favorite activities. Television, recorded music, a good meal, a nice warm house—all of these would be impossible without electricity.

Although it can indirectly cause obesity by way of an electric range, electricity *per se* is not fattening. Neither is it illegal—although it can be installed illegally, as will be discussed later. Electricity is, however, most definitely dangerous, being capable of maiming, killing, and otherwise adversely affecting the human condition.

The capability of electricity to burn and destroy makes most of us pretty nervous, as does the well-publicized existence of electrical codes. And the possibility that we could be violators of these codes is equally frightening.

Actually, these fears are justified. It is a certainty that sticking your hand willy-nilly into any type of live electrical apparatus can be injurious to your health. A code violation *can* burn your house down, although damage to life and limb is far less certain than it would be if you stuck your hand into an entrance panel. These facts *should* make us wary. But healthy fear should not turn us into helpless cowards. A little care and common sense can turn the electrical tigress into a harmless kitten.

Doing your own electrical work will save lots of time and money. Contractors' charges for residential jobs are proportionately far higher than for commercial work. A licensed electrician must bill you much more per unit to install a single switch or outlet than for a whole series of them. And you can be sure that such a job will be very low on his list of priorities. He'll get to you when he can work you in—and that can be a very long time.

Complying with Electrical Codes

Similarly, the mere fact that electrical codes exist should not cause us to tremble with fear. The National Electrical Code (NEC) was formulated for a good and specific reason—to provide guidelines for safe electrical installation. Without it, contractors and electricians could unknowingly wire houses so that they would self-destruct. A do-it-yourselfer could easily endanger himself and his family. The odds are high, however, that you won't wind up in the high-voltage seat at Ossining simply because of a Code violation.

This "code-a-phobia" may be se-

Fig. 1-1. Imagine this kitchen without electricity—no range, no refrigerator, no exhaust fan, dishwasher, toaster, nor many other unseen items. (Westinghouse)

verely exacerbated if the prospective do-it-yourselfer decides to look at a copy of the Code itself. The Code is intended for licensed electricians (and those studying for their license), who are involved in more complex wiring of industrial or business installations. It is by no means a how-to manual for the average homeowner who wants a simple answer.

But this does not mean that the commonsense rules that make up the Code are beyond understanding. As applied to the typical home, they are quite easy to comprehend and follow if explained in nontechnical language. You don't have to understand that "Where a busway is used as a feeder, devices or plug-in connections for tapping off a subfeeder or branch circuits from a busway shall contain the overcurrent devices required for the protection of the subfeeder or branch circuits" (#364-12). I don't, but it doesn't matter. It doesn't apply to me or to you.

You *can* understand that "All junction, pull or outlet boxes shall be accessible without the necessity of removing any part of the building structure, paving or sidewalk" (#370-19). You might not understand completely, at this point, but you get the general picture—boxes must not be placed where you can't get at them.

The point is that you should not be *afraid* of the Code. Essentially, the National Electrical Code is quite simple, as applied to residences, and easily followed if you study and apply the instructions in this book. The *raison d'etre* for the Code is basic and unassailable—to protect you and your family.

As mentioned earlier, the National Electrical Code (NEC) has been formulated for the protection of the homeowner (among others). I have made every attempt in this book to provide instructions consistent with the NEC, but changes are made constantly, and not every contingency can be touched on here. The do-it-yourselfer should make sure that everything he does is in harmony with both the national and local codes.

Many municipalities have their own codes, some of which may be outmoded and more restrictive than the NEC. There is no point in fighting city hall, though, so make sure that whatever you do conforms to local codes as well. It's better than having to tear everything out and start all over again.

Rural, exurban, and some suburban areas may have no codes of their own. In that case, be sure to follow the NEC.

It should go without saying, but in case you haven't heard of it, the Underwriters Laboratory tests all electrical equipment submitted to it for safe design. Never use any material or device that does not bear the UL label.

Fig. 1-2. No matter which way it's written, make sure that all electrical items you use have the approval of the Underwriters Laboratories.

Safety First

There is potential danger in doing your own wiring, but that's like saying water is perilous because you can drown. The intelligent thing for someone who wishes to enter the water is (1) to determine its depth and (2) learn how to swim. The intelligent thing for someone who wants to do his own wiring is to (1) determine what can be done in absolute safety and (2) learn how to do it. You *can* learn how to handle electricity just as you can learn how to swim.

This book assumes one important

thing, often overlooked in other similar books—that the reader, like the writer, is scared to death of electricity. I am. Nevertheless, I have done quite a few electrical projects in my house, and you can do the same in yours. The secret is in knowing how to do them with perfect safety. And it isn't really all that dark a secret, either. The principal rule is simple, and I can't repeat it too often. *Never work around live juice.* Juice, of course, means power.

As long as the circuit breaker has been tripped to "off," or the fuse pulled, you can work with complete safety on the lines served by that circuit. Sure, someone can trip the breaker back on while you're working, or someone can replace the fuse, but that's unlikely, and there are things you can do to prevent this, which we'll go into later. The important thing to remember is that you are—honest—perfectly safe as long as the power is off.

We will show you how to determine which circuit to close down, and to be sure that it is, in fact, out of action before you touch anything remotely dangerous. We will also give you other safety tips to guard against such eventualities as someone replacing the fuse while you're working.

Electricity is indeed dangerous. That's something else we will repeat *ad nauseam.* But there are many *irrational* fears that we will try to eradicate. Here are two common questions that spring from such fears:

Q: *What if I put the wires together wrong?*

A: First of all, you shouldn't, not if you follow my directions. But admitting that none of us follow directions accurately all of the time, and that it is feasible that the direction-giver may not always explain clearly, it is *possible* that you will attach the wrong wires to the wrong terminal. What then? Will the house burn down? No. What will happen is that you will blow a fuse or trip a breaker, or part of the circuit will simply not operate. It is *impossible* for a fire to start just because you have attached the wires incorrectly. (Fires *can* start from other electrical causes, usually frayed insulation or loose connections, but not because you misattached the wires.)

Q: *If I do have a fire, won't the insurance company deny coverage because of do-it-yourself wiring?*

A: No. Check your own fire or homeowners insurance policy. Is there anything in there stating that the company will refuse coverage because you have done your own electrical work? There hasn't been in any policy I've ever seen, and I've seen plenty. If you still don't believe it, call the local claims department, and ask them the same question.

What You Can Do Safely

As long as you are careful, follow instructions, and make sure that you aren't working with live wires, there is a multitude of electrical work that you *can* do around the house. Some jobs are easier than others, but all of them can be done with complete safety as long as you follow the rules.

You can, for example, put in a dimmer-controlled switch for your dining room light. That's not only safe, it's simple. As a matter of fact, you can change any switch or receptacle in the house without fear if you make sure there is no power to the particular circuit you're working on.

More than that, though, you can add to any circuit with complete safety—if you make sure it won't become overloaded. You can put in an entirely new circuit (with the cautions referred to above), and you can put in new lighting, new outlets, a three- or four-way switch—do a hundred other assorted tasks—all without fear. But again, you have to follow the basic safety rules.

Here are some of the basic electrical jobs you can do around the house with complete safety, assuming you follow the elementary rules listed above:

- Replace any switch or outlet.
- Change a lighting fixture, chandelier, and so on.
- Install improved wiring devices, such as a dimmer control for lowering and raising the amount of light in a fixture.
- Replace, or repair, lamp cords, plugs, or any other part of a lamp or appliance that can be disconnected by a plug.
- Add new switches, outlets, fixtures, or similar devices to an existing circuit—if the circuit can handle it.
- Add a complete new circuit, indoor or outdoor, assuming that there is sufficient wattage available at the entrance panel.
- Perform any task involving low-voltage wiring, which includes doorbells, chimes, thermostats, and intercoms.
- Rewire antiquated circuits, or replace any other component of old wiring systems, as long as you do not exceed the present capacity of the circuits.
- Change incandescent lighting to lower energy fluorescent.
- Install built-in track, or indirect lighting, or a luminous ceiling—again, as long as you don't exceed circuit capacity.
- Convert most switches to three-way or four-way, so that a lighting fixture can be turned on or off at several points.
- Wire an addition, finished basement, attic, or garage.
- Perform numerous electrical improvements or repairs not directly related to wiring, such as appliance trouble-shooting or planning better room lighting.

What You Should Not Do Under Any Circumstances

As mentioned above, the most important rule in all electrical work, unless you're a professional and sometimes have to, is not to work when the current is on.

Most how-to wiring books explain how to connect a new circuit to the service (exterior) side of the entrance panel. I'll do that, too, but I will also advise you not to make the final hookup to the panel yourself. This *can* be dangerous. For the sake of completeness, I'll give you the details, but I urge you to hire an electrician, have him check out the circuit, and make the hookup to the panel for you. There are lots of advantages to this, including the assurance that you've done the preliminary work right, but the most important is that you won't run the risk of touching a live portion of the panel when you install a fuse or breaker.

You may also find directions for adding on a subpanel or auxiliary fuse box in some books. I'm not even going to discuss this. I think it's a dangerous thing to do, because that also involves working with the entrance panel.

Equally dangerous for the do-it-yourselfer are attempts to increase "housepower" or add to the capacity of the entrance panel. If you need increased capacity, call a qualified electrician or the lighting company.

1. The most important rule is always make sure that the current is off before touching any electrical device. Never assume a circuit's dead just because you threw a switch or breaker, as these devices sometimes short-circuit internally. Verify that the circuit is dead, either by testing or by trying to turn on something on the circuit (e.g., a light bulb).

2. Never work with wet hands, feet, or other part of your body. Electricity does funny things when it's damp, so avoid damp or wet areas, or wear rubber gloves and/or boots when working where it's wet.

3. Whenever possible, work with only one hand, keeping the other in your pocket.

4. Don't touch anything in the entrance panel or fuse box except for the breaker or fuse you are disconnecting. As an extra precaution, stand on a dry wooden board whenever you touch the main panel, particularly if there is dampness around.

5. Never overload a circuit. Ways of determining if and when a circuit is overloaded are discussed on pp. 61–63.

6. Never substitute a fuse or breaker with one of higher amperage. A fuse prevents the wires from overheating because the lines are carrying more current than they were designed to carry. If you replace a 15-amp fuse with a 20-amp fuse, the wires will get too hot and melt the insulation, causing a dangerous situation and possibly a fire. More on this in Chapter 2.

7. Wiring exposed to the elements or dampness should be doubly-protected by a ground-fault interrupter (GFI). This provides an extra measure of safety for both humans and animals by cutting off the current within 1/40th of a second. A fuse will blow or a breaker will trip in the same situation, but the GFI does it faster and eliminates any chance of dangerous contact with live circuits. (See pp. 112–113 for details.)

8. Avoid loose connections, nicked wires, and aluminum wiring.

What You Can Do at Some Risk

To be consistent with the theme of this book, this section should perhaps be incorporated into the previous one—jobs that you shouldn't tackle at all. In truth, however, there are several wiring tasks that can be done safely if you exercise a reasonable amount of caution and common sense. They are not recommended for the spanking-new amateur.

If you are at all faint-hearted, don't go near the entrance panel except to pull or replace a fuse or reset a circuit breaker. The valiant (who never taste of death but once) can give it a whirl, but only with extreme caution.

Actually, there are certain types of entrance panels that can be completely shut down by pulling the "main" switch. But there are others that retain some current to the range and other circuits. Therefore, there may be some power into which a stray hand may wander, even with the main breaker thrown. It is not always easy to determine which type of panel you have, and that is why it is better, for most of us, to wire up our circuits all the way up *to* the entrance panel, and have a professional actually make the connection for us. This still saves a lot of money, while at the same time preventing the clumsy among us from self-immolation.

Fundamentals of Electricity

Electricity is a curious phenomenon. It is the exact opposite of the weather. Nobody complains about electricity, but everybody does something about it. Flip a switch and a light goes on. Plug in an iron, it heats. Turn on a motor, it runs. Electricity brings us entertainment via television, cleans the dishes, brushes our teeth, compacts garbage, activates burglar alarms, blends cocktails, and warms coffee.

Unless you live on a desert island, you use electricity every day in more ways than any of us realize. Not until there is a power failure do we appreciate how dependent we are on electricity. But do we know what it *is*? Few people can tell us in a way that makes it easy for us to understand. It has to do with ions, electrons, magnetism, and all sorts of complicated theory, which, frankly, I don't understand.

However, it really isn't necessary to understand the complex physical laws that explain what electricity is. What you can and should know is *how* it works in your home. To explain this, I'll use a simple analogy.

How Electricity Works in Your Home

Most of us understand how a plumbing system works. Odd though it may be, the electrical system works in much the same way. Substitute wiring for pipes, and electrical current for water, and you can get some idea of how electricity works.

First, you know that no water flows in the pipes until a faucet is turned on. The same applies to electricity. The capacity (water) is there, but it doesn't go anywhere. Turn on a water faucet, and water is brought in under pressure from the water plant. It flows through the water main through smaller pipes into the faucet of the bathtub or sink. In the same way, when you turn on a switch, current is brought in from the power plant through transmission lines into the entrance panel, then through the electrical circuit (pipes) into the outlet (faucet).

Rate, Volume, and Resistance

The rate of water flow is determined by several factors, including water pressure. Water pressure is roughly analogous to *voltage*, in that both are determined by type and size of the power plant. Water flow inside the home depends on the size of the pipes—the bigger the pipe, the more water it can deliver. The "pipes" of the electrical system are the wires themselves. A bigger wire can carry more current, smaller wires less current. This rate of water flow past any given point is analogous, electrically, to *amperage*.

The actual volume of water that is delivered from the tap depends on both the water pressure and the size of the pipes. The electrical equivalent of this volume is *wattage,* or the amount of electrical energy consumed by the light, appliance, or other electrical convenience.

Another factor that affects wattage is *resistance,* a term not readily adaptable to the water system analogy. The most apt comparison is to old pipes that are corroded or partially blocked.

Overload

You can see how wattage (water volume at the tap) is increased or decreased by the voltage (pressure) at the source, and by the amperage (water flow) past a given point, which is in turn determined by the size of the wire (pipes). But the water flow analogy does not properly explain what happens when the wattage is increased beyond the danger point. If you turn on all the water taps in the house at a given time, you get only a trickle through all of them. Electrical flow, on the other hand, is constant. The more "taps" you turn on, the more electricity is delivered. The flow, or amperage, does not diminish, as it would in the individual lines of the plumbing system.

As a result of the excess flow, the strain on the circuit becomes too much to handle. If the home is wired correctly, the entire circuit shuts off by tripping a circuit breaker or blowing a fuse. But the circuit breaker can be reset; the fuse can be replaced. The homeowner, however, must determine what caused the overload. The culprit in a "blown" or overloaded electrical circuit is excess wattage, most often caused by too many heat-producing appliances, which demand more energy than the circuit can deliver.

Useful Terminology

To repeat, it is useful, but not necessarily vital, to understand what electricity is. It is certainly helpful to understand how electricity works, but if the analogy in the previous section didn't prove helpful, don't despair. All you really have to know are a few terms and definitions. If you familiarize yourself with the following terms, you should be able to do most of the tasks in this book.

Ampere: The unit used to measure the amount of electric current passing through a given point on a circuit each second, abbreviated amp.

Circuit: The wire or wires through which electricity flows from the service panel to where it is used, and back.

Current: Electric movement or force (juice).

Circuit Breaker: A device that trips when a circuit is overloaded. The breaker thus prevents wires of the circuit from becoming overheated due to too much current, which can result in damage to other devices on the circuit, burned insulation, and possibly fire. Breakers can be reset, once the problem is eliminated.

Fuse: A device that performs the same function as a circuit breaker, but that must be replaced by a new fuse rather than reset; found mostly in older homes.

Outlet: An electrical "tap" that supplies power through a plug-in wire to a lamp, appliance, and so on. Sometimes called a receptacle.

Overload: A condition that results when a circuit is forced to carry more wattage than it can handle. Produces a "blown" fuse or tripped breaker when system is properly wired.

Resistance: The loss of energy, most often in the form of heat, caused by a material used to carry an electric charge; measured in ohms.

Volt: The unit used to measure the pressure of electromotive force (EMF) that gets an electric current moving and keeps it active.

Watt: The unit of power that indicates the rate at which a device uses energy determined by multiplying amps by volts.

How Electricity Enters a Home

Electricity is "generated" or manufactured by some source of energy at your local utility. The source may be falling water, as at Niagara Falls; coal, as in many Western U.S. power plants; oil, as in much of the world; nuclear fission, or even windmills, solar panels, or other exotic sources, such as geothermal power from deep inside the earth.

Transformer and Service Lines

The energy, whatever its origin, powers huge turbines that excite the electronically charged particles called electricity. This high-voltage electricity is then discharged via power lines to a transformer outside your house. The transformer tames the higher voltage to a lower one, usually 120 volts, which

Fig. 2-1. Small amounts of electricity can be generated by a person riding a bike. If this woman pedals fast enough, she can light the 150-watt bulb. Continuous pedaling for 8 hours would yield 1.2 kilowatt hours of electricity—worth about five cents.

is then carried via further service lines to the electric meter, where it is measured for use, and into the entrance panel.

Two- and Three-Wire Service

Most modern homes have three wires leading to the entrance panel. Two of them are hot wires, with a potential of 230 volts. The third is a neutral, or ground, wire, completing the circuit back to the power source. Between each hot wire and the ground wire, a potential of 120 volts exists. In older homes, there is usually only one hot wire. This is called two-wire service as opposed to the now-prevalent three-wire service. The extra wire allows the use of high-wattage appliances such as electric dryers, ranges, and central air-conditioning. For these appliances, the voltage is doubled by using power from two 120–volt incoming wires. This is often called 220, although the actual voltage is usually 240 (sometimes 230).

Home Circuitry

Circuits

At the entrance panel, the incoming power is subdivided into "circuits," by which the electricity is distributed to different areas of the house. Each circuit is tapped into the panel through breakers or fuses. These devices protect against overload. If the power demand on any circuit exceeds the capacity of the wiring (too much amperage), the entire circuit shuts down, as explained above, in order to prevent damage to the wiring or the appliances connected to it.

An individual circuit can consist of just one large appliance, such as a range. If you look into your own entrance panel (not recommended until you've read further), you will probably find a large breaker marked range, which consists of two smaller breakers joined together, each marked 40. This means that the line consists of two linked 40-amp circuits, with a total current capacity of 80 amps. You may also find a separate dryer circuit, usually consisting of two 30-amp breakers with a total capacity of 60 amps.

The other fuses or breakers in your home are most likely 15 or 20 amperes, usually both. The 15-amp circuits are probably lighting circuits leading to the fixtures and outlets in the living room, dining room, or other quieter areas of your home. Often, the circuit may traverse several rooms, such as two or three bedrooms.

You will probably find that there are several 20-amp circuits that lead to the kitchen, the most energy-demanding room. Two or more 20-amp circuits may serve the kitchen exclusively, with another that perhaps also includes the basement or workshop area. Since the appliances often used in the kitchen consume more power (higher wattages), they need larger capacity circuits. The same applies to many power tools used in basements and workshops.

Current Capacity

Your home circuitry has been designed by experts who have predetermined the capacity of each of the lines. Often, there is the capacity for extra outlets on each line, but it is important to determine exactly what type of wattage may be expected if you plan to expand that circuit. This topic will be discussed in detail in Chapter 6, but the important thing to remember at this point is that each breaker or fuse has also been designed to "fail" if and when the capacity of the circuit is exceeded.

Sudden surges, a faulty appliance, or other temporary problems may lead to an occasional blown fuse or tripped breaker. Once the problem is corrected, there should rarely be a repeat. (Frequent blown fuses mean either faulty wiring or an overloaded circuit. Often, however, when a fuse blows repeatedly, the homeowner may decide that a larger capacity fuse will correct the problem.

Never replace a blown fuse with one of a higher amperage. As noted in Chapter 1, this is an important, fundamental rule of home wiring. It also applies to circuit breakers, although, fortunately, it is more difficult and dangerous to replace a circuit breaker. It is deceptively easy to replace a 15-amp fuse with a 20-amp one. But you must understand that the fuse is rated for a certain size of wiring as well as a maximum amperage. A 15-amp circuit will most likely be composed of #14 wiring, which is a little thinner than #12 wiring. The difference in diameter is small to the naked eye, but crucial as far as heat generation is concerned. If 20 amperes are allowed to flow through #14 wire for an extended time, the wires will overheat, burning the covering insulation, thus exposing the bare, hot metal and eventually starting a disastrous fire. Any appliance connected to such a circuit can burn out as well, because more power may flow into it than is supposed to.

To repeat then, because it is so important, never replace any fuse with one of higher amperage.

Completing a Circuit

Electrical current cannot flow through a wire that is "open" at one end, that is, not connected to anything. For electrical current to flow through a wire, the wire must be part of a complete circuit, from the source, through the wiring, and back to the source. For example, a wire that has come loose from its terminal will open the circuit, and no electricity will flow to the outlet. In most cases the entire circuit goes dead (although not necessarily, if a switch or short is involved).

Whenever you find that a circuit is not operating properly but that the fuse or breaker is okay, you can assume that there is a missing link somewhere in the circuit. If this happens after you have installed your own wiring, you have probably hooked up the wires to the wrong terminals, or have left something out some place. (This gets more complicated where there are splices, switches, etc., involved, but more on that later.)

What you should remember at this point is that you must be able to measure the line voltage everywhere on the circuit, even when no load is actually connected to the circuit. As soon as the continuous flow is interrupted, the current ceases to exist. As a matter of fact, this is exactly what happens when a breaker is tripped. The continuous flow is cut off at the source, the breaker. With the breaker off or the fuse pulled, there is no electrical current in that entire circuit.

To a lesser extent, the same is true when you turn a switch to off. What the switch does is cut off the flow of electricity to the fixture. It is *possible* (but definitely not recommended) to work on the fixture itself with the switch off, because there is no power at that point. Always be safe and cut off the circuit at the source.

Short Circuits

There is one other way of blowing a fuse or tripping a circuit breaker—a "short circuit." What happens here is that a bare wire touches another bare wire, or a metal part that is grounded. In this instance, current flows away from the normal channels into the other wire or the grounded metal. Lacking any resistance along this bypath, the current becomes very high, thus tripping the breaker or blowing the fuse.

Shorted or otherwise inoperable circuits must be "cured," of course, and the procedures for tracking down the "disease" are covered elsewhere in this book (see p. 66). What you should grasp at this point is that electricity is distributed to all parts of the house via certain definite pathways known as circuits, and that the capacity of these circuits and their components have been precisely calculated at the time of their installation to best serve the needs of your home without being the cause of any dangerous situations, such as shock or fire. It is possible to extend, reroute, or otherwise alter a circuit, but these changes must be carefully planned in advance. It is also possible, in many cases, to add entire new circuits, if there is enough current capacity at the entrance panel.

Housepower

A word frequently used by utilities and electrical contractors is "housepower." If your home was built within the past 10–15 years, you probably don't have to worry about housepower, but it can be a real problem in an older home, or to someone who would like to switch from natural gas to electric power for heating or cooking.

Definition

"Housepower" is the "service" (jargon more often used by electricians themselves), meaning the conductors and equipment for delivering energy from the supply company to your home. If your housepower is low, a contractor will advise you to install a new or expanded service. What he means is that you will have to increase the ampere capacity of your home. This usually involves more or heavier lines coming into your meter, or it could mean increasing the power with a new entrance panel, or, perhaps, by adding an auxiliary subpanel. It may also be necessary to install heavier wiring capable of carrying the increased current loads.

The problem originates because earlier homes were not designed for the appliance-mad culture of today. Older homes may have had only a 30- or 50-amp capacity. A home with a 100-amp entrance panel at that time had a generous capacity indeed. Now 150 amps is considered adequate, and 200 amps not overly generous. A lot, of course, depends on whether you also have access to natural gas or alternate types of power. With electric heat, you'll need a considerably greater capacity than most.

How to Increase Housepower

In any case, you know that your housepower is low when all the circuits are fully utilized and cannot accommodate any more appliances, or fixtures. If fuses blow often, or you find yourself using a lot of extension cords to accommodate lamps or appliances, you should seriously consider increasing the ampere capacity of your home.

This is no job for the do-it-yourselfer. A licensed electrician should do the job from start to finish. It's tricky, dangerous work, so don't even think of trying it. What you can do is look into the entrance panel and determine what the listed current capacity is. It should be listed on the inside of the door—or somewhere nearby.

If you cannot find this information, look at all the fuses or breakers, adding their listed ampere capacities together, so as to arrive at a total ampere capacity for the panel. This should give you at least a rough idea.

If you have fuses instead of breakers, do the same thing, but the mere presence of the fuses is a good clue to undercapacity. The service is undoubtedly old, and unless extra capacity has been added since the house was built, chances are good that you can use more housepower.

All of the above clues, however, are just that—indications, not a final answer. An entrance panel may be listed at 125 amps, but the wiring on the house side may be too small to carry this load. Determining the exact housepower of your home can be quite complex. If necessary, ask your utility or a licensed electrician to determine capacity for you.

3

Tools and Materials

Fortunately for the do-it-yourselfer who wants to tackle an occasional electrical job, a few common tools—a screwdriver, pliers, and a penknife—are all that are really essential. If you are planning more than just a switch replacement or lamp repair, however, you will want a few more specialized tools. They make the job easier, faster, and better. Depending on what the job is, it is best to accumulate the essential tools first, then pick up others as you need—and can afford—them.

For some simple repair jobs, like reattaching a loose wire, you won't need any materials at all. Minor add-on jobs, such as an extra outlet or two, will require a few feet of nonmetallic, sheathed cable (popularly called Romex), boxes, receptacles, and some solderless connectors (wire nuts). Obviously, more complicated jobs need more specialized tools.

Tools

Basic Tools

If you're getting into anything heavier than an infrequent repair or add-on job, it is wise to check your tool inventory. You may already have some of the tools shown in Figures 3-1 and 3-2. Others can be purchased as needed.

Electrician's Screwdriver

Similar to standard screwdriver, with a long, slender blade and a tip the same diameter as the shank. The blade fits snugly into the screwheads of most electrical connections and the shape makes it easy to work in deep-set and difficult to reach spots. The handle should be insulated to protect against shock.

Wire Stripper

This simplifies stripping off insulation. You can do the same thing with a penknife, but the stripper has settings for each wire size so that it takes the insulation quickly without damaging the wire. Strippers come with plastic cushion grips. The one shown in Figure 3-1B cuts and strips 12–24 gauge wire.

Thin-Nosed Pliers

These have narrow, tapered jaws, making it easy to bend wire around terminals and to reach into tight corners. The better ones have insulated, cushion grips.

Cable Ripper

Used to cut through outer sheathing from nonmetallic cable (Romex). The

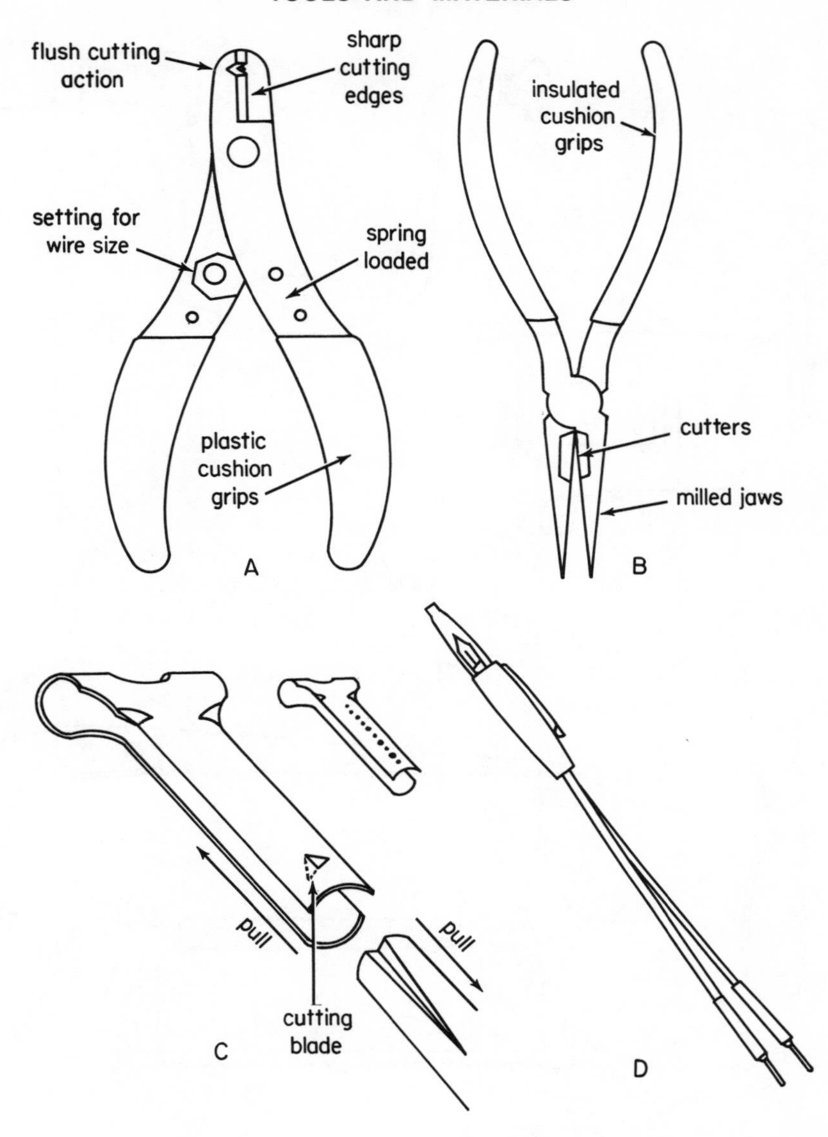

Fig. 3-1. (A) Wire stripper, (B) Thin-nosed pliers, (C) Cable ripper, (D) Tester.

needle-sharp cutting blade rips cable easily without damage to the inner conductor. The one in Figure 3-1C measures 14 to 6 gauge cable.

Claw Hammer

A 16-ounce hammer for driving staples, nailing up outlet boxes, fastening hangers, and similar work.

Folding Rule (6 Reg. Foot)

The best tool for measuring wall openings for boxes, wire lengths, and so on.

Tester

A tool to safely detect whether or not voltage is present at points along a circuit. Its main purpose is to check

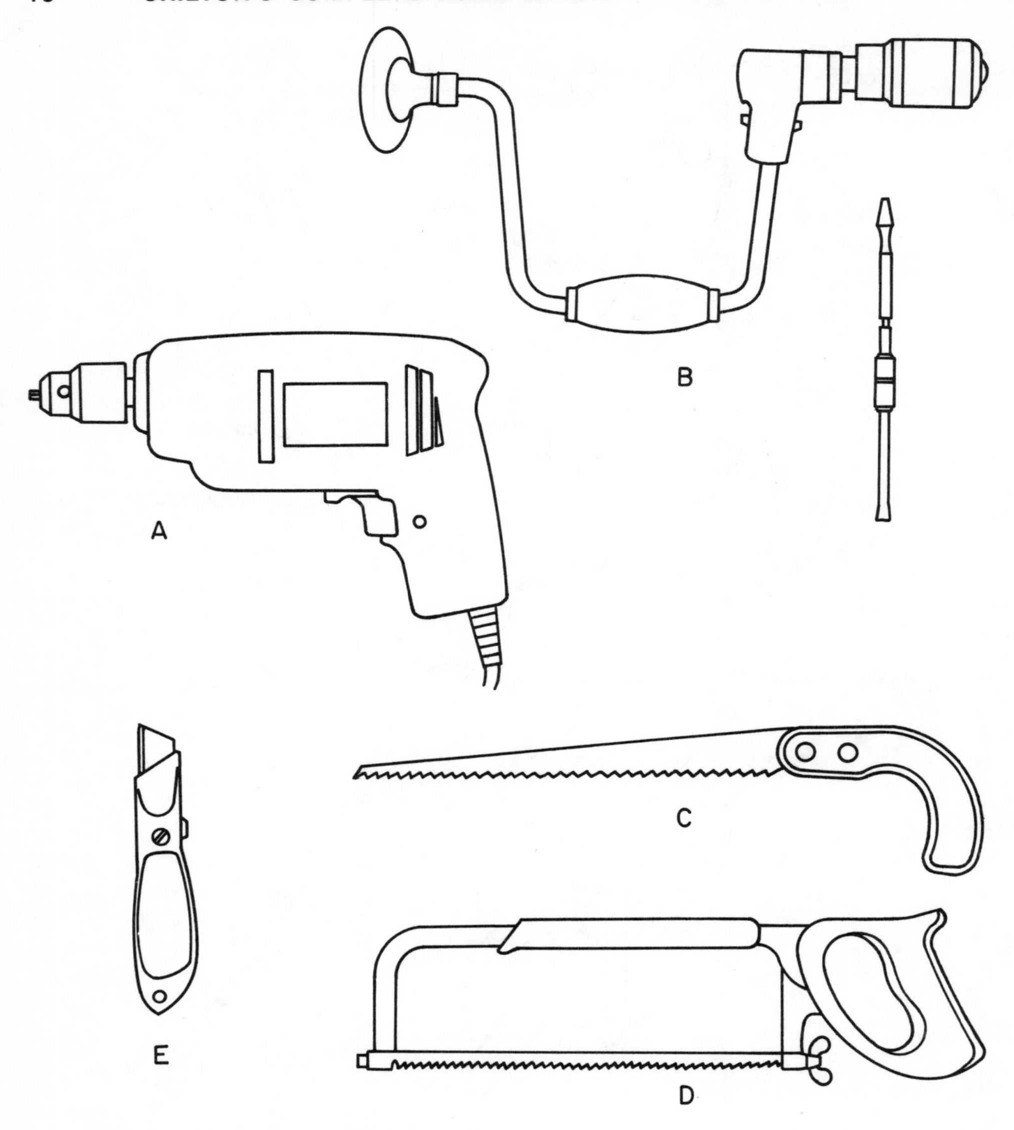

Fig. 3-2. (A) drill; (B) brace and bit; (C) keyhole saw; (D) hacksaw; (E) utility knife.

that the current is off before you begin to work. It is also used with the power on to check for proper grounding. Although there are some complicated types, a simple one, with two prongs and a small light, is sufficient for most jobs.

Drill

An electric drill is best, of course, but there may be times, in electrical work, when there is no electricity available. In that case, a brace and bit (with ratchet) comes in handy. For such work, a ⅝-inch bit is the most useful. Long extension bits are a necessity for running cable through existing structures.

Keyhole Saw

Handy for cutting out box locations in gypsum wallboard and paneling. A

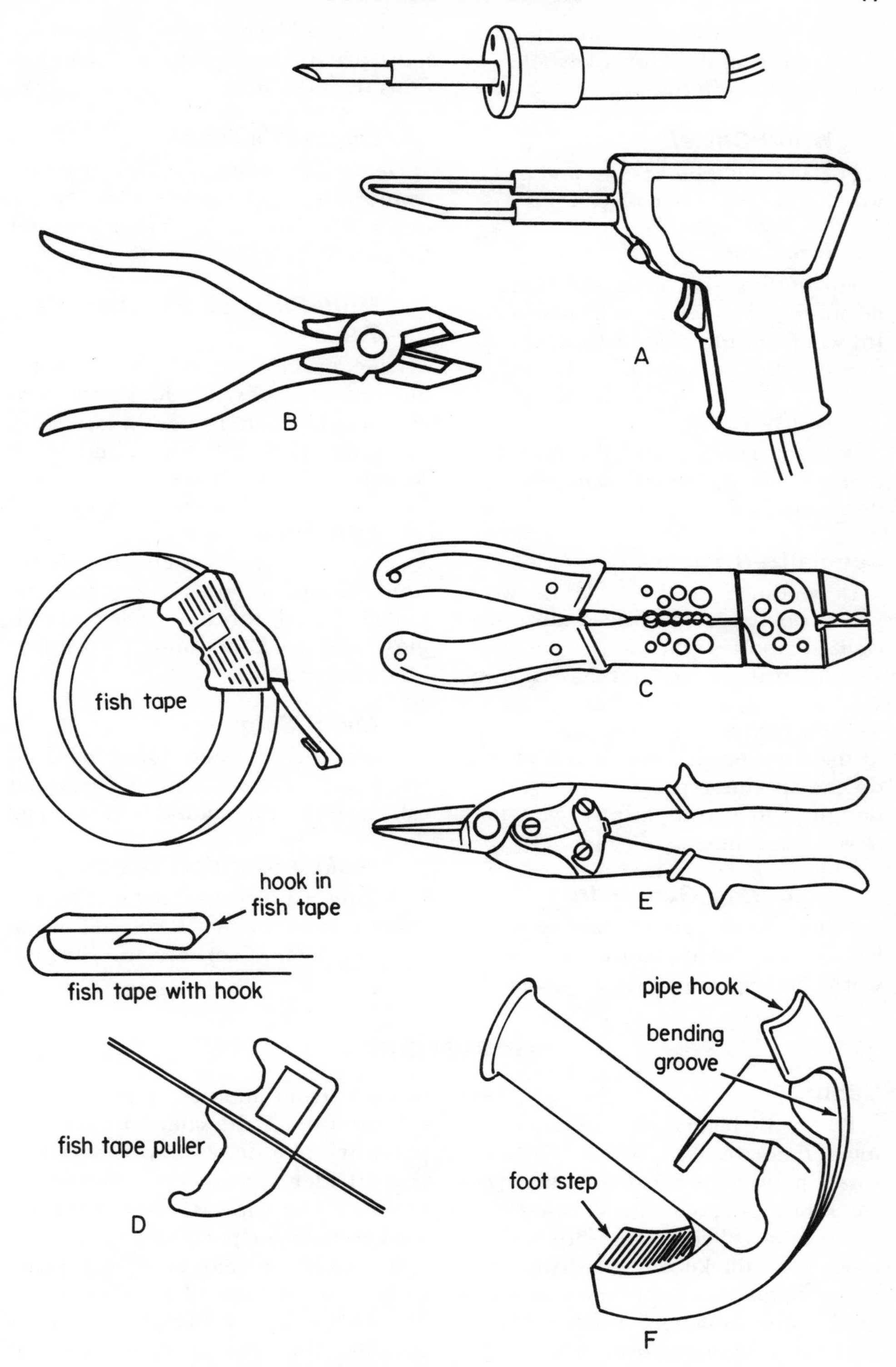

Fig. 3-3. (A) Soldering iron and soldering gun; (B) lineman's pliers; (C) multipurpose electrician's tools; (D) fishtape, hook and tape puller; (E) metal snips; (F) conduit bender (hickey).

saber saw does the same job faster, but you'll also find uses for a keyhole saw when working in tight places.

Wood Chisel

For cutting notches in wall studs, wood lath, and other places.

Hacksaw

For cutting the metal sheathing of armored cable (Type AC). Also helpful when cutting metal lath. Use a 32-tooth blade if cutting thin wall conduit.

Utility Knife

A razor knife, about the only tool that can cut the tough plastic on underground cable.

Specialized Tools

Depending on the type of work you're doing, you may need more specialized tools right away, but they should usually be purchased after you've gone through basic training with the others mentioned above. Most of them are needed only if you plan to do the specific type of work they are designed to accomplish, so you may never need some at all (Fig. 3-3).

Soldering Gun or Iron

Not generally used in wiring *per se,* but useful for appliances, electronic work, and splicing (which should be rare). A soldering gun is generally more useful for splices and connections than the iron.

Lineman's Pliers

Heavy-duty pliers used for gripping connectors, cable, and other larger items. Also has side jaws that are used for cutting cable.

Multi-Purpose Electrician's Tools

Pliers in different varieties, combining a number of operations into a single unit. The better ones do crimping, stripping, cutting, holding, and other common electrical tasks.

Fish Tape

A necessity when working in existing construction, where wires must be pulled (fished) through walls and ceilings. Also used for conduit work (see Materials section).

Metal Snips

Aviation type is best. Often used instead of a hacksaw for cutting armored cable. The handle should be insulated.

Hickey (Conduit Bender)

A long-handled tool for getting the correct radii in metal conduit when working around corners and obstructions.

Materials

Cable

By far the biggest item in home wiring is the wire itself, which comes encased in various insulating materials for safety purposes. So encased, it is technically called cable, although to most it is still known as wiring. The wires are made of copper, which is covered with insulation, usually plastic, and the insulation-covered wires are then wrapped in an outside covering as described below. It is the outside wrapping that distinguishes one type of cable from another. The wires inside are usually the same, copper with polyvinyl chloride (PVC) plastic insulation. (Older wires were covered by rubber.) The various outer coverings are specified for particular usages.

It should be said here that many codes allow the use of aluminum wiring. Others are in the process of rescinding this allowance, however, if they haven't already. Even if your local code allows aluminum wiring, don't use it. Studies have shown that there is too much danger of heat build-up and loss of contact at crucial points in the system.

If your home now contains aluminum wiring, you should have the connections "pig-tailed" as soon as possible to avoid a potentially fatal fire. Pig-tailing is described on page 17. If you haven't the time or know-how to do this yourself, don't wait for either. Have an electrical contractor do it for you.

Most cable now comes with a separate ground wire in addition to the black and white insulated wires. This extra wire is a safety ground, and is eventually linked up to an underground water pipe or some other type of ground terminal, to guarantee that a fuse will blow or a circuit breaker will trip if there is a short circuit. The white wire is a neutral wire that carries current back to the panel. Cable that contains an extra ground wire is marked "with ground" or W/G. The ground wire is either bare and copper-colored, or covered with thin green insulation.

If you are extending an older circuit that has no ground wire, you should still buy cable with a ground wire, even though the ground cannot extend back to the entrance panel. It can still be grounded to the box to protect against a short circuit. Also, you may as well hook it up correctly (see p. 57), in the event that someday you may replace the older groundless cable, in which case you can connect the ground wire to the replacement.

There are numerous types of cable, most of which are used in industrial or commercial applications. The types that you will most frequently encounter in the home are the following.

Nonmetallic Sheathed Cable

Popularly called Romex, this type of cable is the kind most often used in the home (Fig. 3-4). Because of its soft cov-

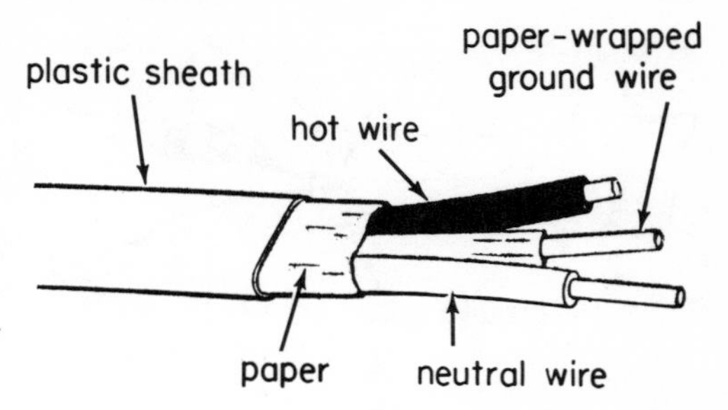

Fig. 3-4. Nonmetallic, sheathed cable (Romex).

ering, it is relatively easy to work with and can be used in most home installations, but not in places where the wiring might be exposed to wetness or damage. Some local electrical codes do not allow the use of Romex in any location—a silly and costly restriction that is, nevertheless, a boon to contractors. There are several different subtypes of nonmetallic cable.

Type NM: This is the type most frequently used in home installations. It can be used in concealed or exposed locations, and can be "fished" through existing walls. Because of its soft covering, it cannot be used outdoors, underground, or in damp locations.

Type NMC: This moisture-resistant cable can be used outdoors and in damp locations, but not underground.

Type UF: A tough-skinned cable that can be used underground.

Type SE: SE stands for Service Entrance, which is its primary use. You could possibly use this thick cable for the range or dryer.

Flexible Armored Cable

This cable, which is usually called BX, is called AC by the NEC (Fig. 3-5). The wires are covered by galvanized steel, wound in spiral fashion so that it will bend. A homeowner should use this where there is a chance of damage from tools or nails that may be driven into the wall. It cannot be used in wet locations because the steel may cor-

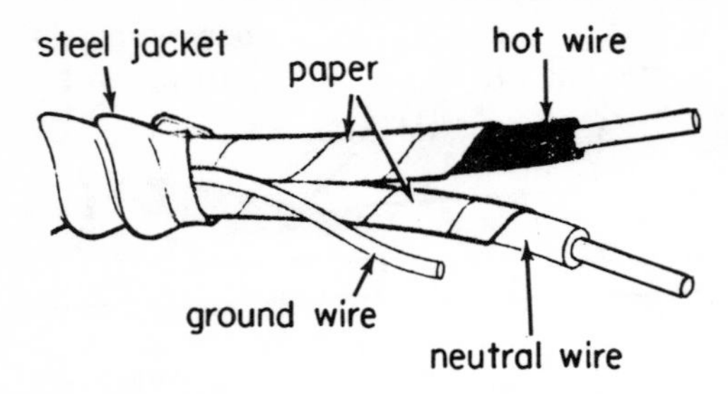

Fig. 3-5. Flexible armored cable (type BX or AC).

rode. It is more difficult to work with than nonmetallic cable. Some local codes still require AC cable indoors, but this is a needless precaution for most installations.

Conduit

Not a cable at all, but a metal (sometimes plastic) sheath in which wiring is run. Conduit is used in many hazardous locations and outdoors when UF cable is exposed. Conduit comes in two types, thin-walled or rigid steel, with the latter preferred for outdoors. Polyvinyl chloride plastic can also be used, but is sometimes prohibited by Local Codes. PVC conduit may be prohibited by a local code because it gives off noxious fumes when it burns. For the same reason, Romex cable is prohibited beyond three stories of a building. Special, unsheathed wiring (Type TW) can be pulled through the conduit after it is installed, but there is little need for this in home wiring. The only time you might use conduit is when underground UF cable runs above ground (see page 112). Rigid conduit is preferred here, so that you will find little use for thin-walled conduit (EMG) unless mandated by local codes. Conduit is expensive and difficult to work with. It also requires special connectors. Thin-walled conduit can be bent with a hickey (p. 78). The various materials used for conduit, as well as methods of installation, are described in this section.

In addition to type, cable is also designated by size. The size refers to the diameter and number of wires only. Thus, 14–2 wiring contains two #14 wires (not 14 #2 wires, as you might expect). Size comes first. Sizes are those used by the American Wire Gauge (AWG) system.

The gauge (number or #) indicates rather minuscule differences in diameter (Fig. 3-6). Though the difference in wire diameters seems very small (only .017 of an inch between #12 and #14, for example), there is a significant difference in the capacity of the wiring. The current-carrying capacity of a wire strongly depends on its cross-sectional area, which increases as the square of diameter. Thus #12 wire (0.0808-in diameter) has a cross-sectional area 1.6 times greater than that of #14 wire (0.0641-in. diameter). The #14 wire, in given conditions, is rated to carry 17 amps continuously, while the #12 wire under the same conditions is rated to carry 23 amps.

Wire sizes commonly used for home wiring can vary from #0 to #18, with most circuits using #14 or #12. (The smaller the number, the thicker the wiring.) The only time you would run into #0, for example, would be at the service, where it, or perhaps #00 (also called 2/0) or #000 (also called 3/0), is used to carry current to the entrance panel. Smaller #16 and #18 wiring is used only for doorbells, thermostats, intercoms, or other low-voltage wiring stepped down by a transformer. Home wiring generally makes use of single-strand wire; "single-strand" means that each length of cabling is com-

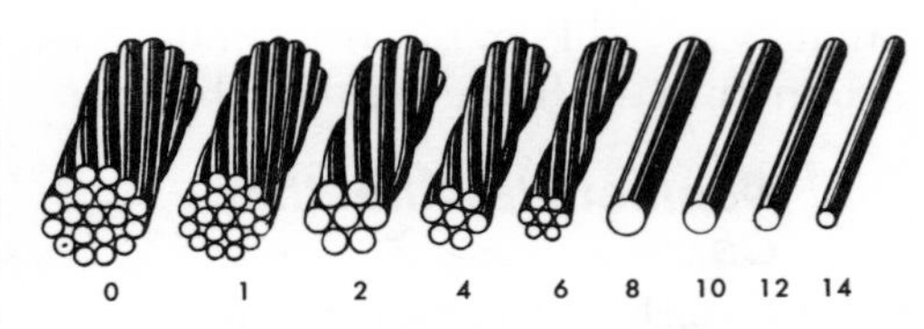

Fig. 3-6. Sample cable sizes, #0 to #14.

posed of an insulated length of a single copper wire. Single-strand wire becomes less flexible as its gauge becomes heavier. Thus any wire heavier than #6 must be multistranded, since the heavier wire will be too stiff to bend. Multistranding means that the wire is composed of many fine strands, and the gauge of a multistranded wire is taken to mean the effective size of the wire bundle considered as a whole. Electric cords, and similar wires that are subject to much bending—lamp cords, appliance cords, and so on—are always multistranded, even though they are usually in the #14 to #18 capacity range.

If you are extending a circuit in your home, always use the same size wiring that is already in use on that circuit (see pp. 71–72). When installing an entire new circuit, there are several factors to be considered in deciding which size wiring to use (see Chapter 6).

The type of wiring can be varied within the same circuit as long as the *size* remains the same. If, for example, you use 14/2 UF underground cable outside, you can switch to NM or AC once you get inside the house, but it must be 14/2.

Boxes

Boxes are an integral part of any electrical circuit (Fig. 3-7). All connections, splices, or any other interruptions of continuous cable must not remain exposed or allowed to hang loose; they must be encased within a box. As implied by the name, a box is a holder for switches, receptacles, and other wiring devices. It is usually rec-

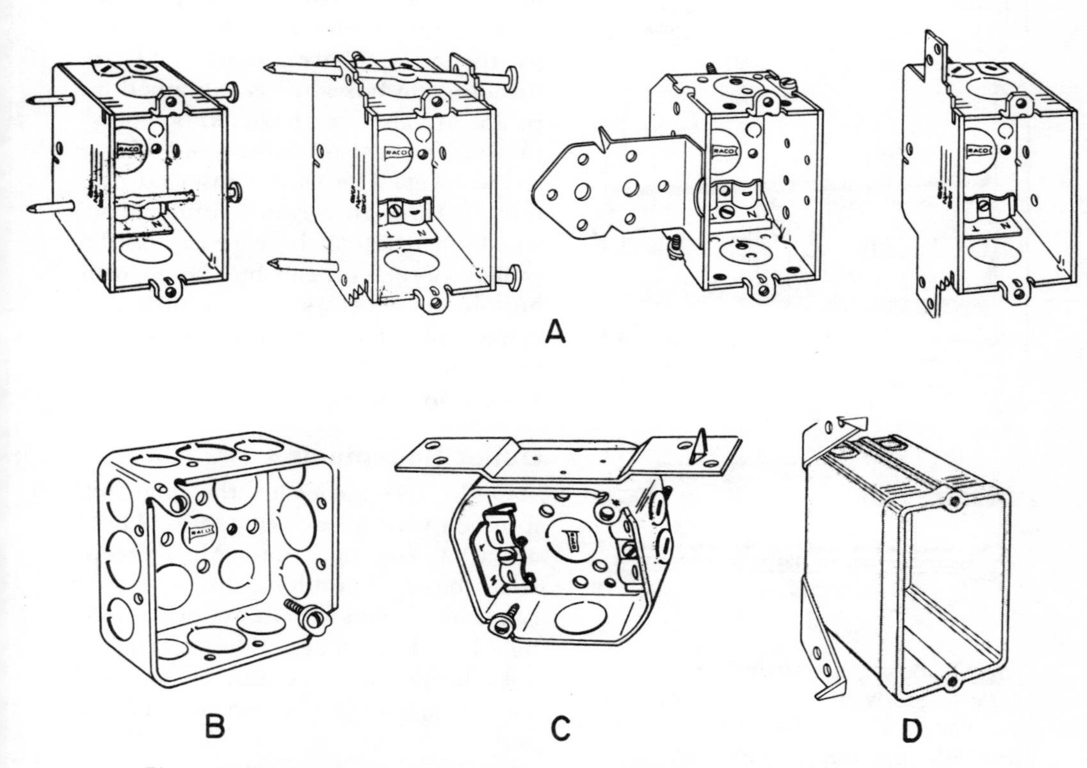

Fig. 3-7. Electrical boxes come in many shapes and sizes. (A) Rectangular boxes, also known as switch or Gem boxes. Note each is attached to framing by a different method. (B) A square "handy" box (C) A ceiling box. (D) A plastic box.

tangular, but can also be square or octagonal. Rectangular boxes are used for most walls, octagonal boxes are usually found in ceilings. Large, square boxes are used for 240-volt installations, such as ranges and dryers. Smaller square boxes are usually used for junctions. Shallow boxes are also available for thin walls or other narrow locations.

Boxes are attached to the house framing by nailing through one of their many holes. Some boxes have plates welded to them or other attachments that make them easier to nail to exposed framing. Most boxes have mounting "ears" at top and bottom, if you can find something to attach the ears to. "Madison" clips are usually required to attach boxes to gypsum wallboard, although some boxes have built-in mounting devices for this (see page 35).

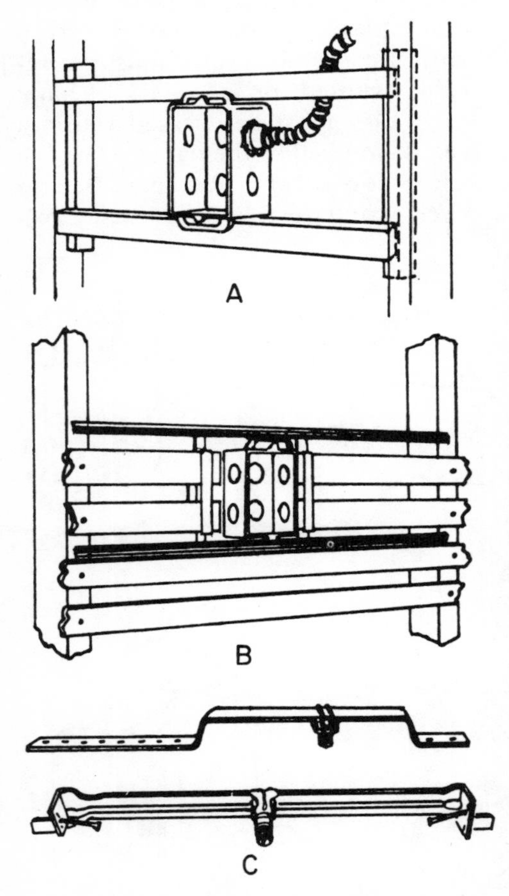

Fig. 3-8. Boxes are usually attached directly to studs or joists, but sometimes you want a fixture located in a certain spot between framing members. In that case, use either (A) wood cleats, or (B) metal mounting straps. (C) Bar hangers are available at electrical supply houses and some hardware dealers.

Special hangers or straps are sometimes employed to attach boxes to exposed framing (Fig. 3-8). These are usually used when you want to mount a ceiling light in a certain spot where there is no framing to hold the box. Boxes with removable sides are used for "ganging" (see p. 86).

Although most boxes are metal, plastic or other nonmetallic boxes are often used in damp locations. These may not be approved by local codes because they come with built-in holes for the cable. There are also NEC restrictions on box size, as explained on p. 85. Metal boxes have "knock-outs" that are tapped or twisted out with a screwdriver. The wire is inserted into the box through various "connectors," which ensure that the wire can not be pulled loose. You can buy boxes with built-in connectors for nonmetallic cable. Special weatherproof boxes, as illustrated on p. 109, are required for outdoor work.

Outlet Receptacles

Outlet receptacles are the final step in much wiring work (Fig. 3-9). These are what you plug into. Most receptacles have two outlets, although they can be equipped with more, or be combined with switches. The receptacle goes inside the box and is attached with accompanying screws into the top and bottom of the box. The outlet is then covered with a plate screw-fastened to it through a hole provided in the center. All new outlets have an extra hole for a three-prong adapter

Fig. 3-9. Receptacles. (Left) two outlets; (right) single outlet. Both accept three-prong plugs.

plug that connects with the ground wire. This adapter plug protects against serious shock. It is a must for power tools.

Receptacles are usually connected to the wiring by screws provided on the sides, as described on pages 35–37. Some outlets have push-in terminals, so that you just have to push the wire into the terminal aperture. The wire is gripped by a spring, which can be released by inserting a screwdriver into the release aperture next to each wire.

Switches

Switches are also inserted into previously installed boxes (Fig. 3-10). They are hooked up in much the same way as receptacles. Plates covering them are attached by two screws at top and bottom.

Three-way and four-way switches allow you to control lights or other electrical devices from two or more locations (three-way means two locations, four-way is three locations, etc.). They are a little tricky to install and change. (See pages 54–56 for details.)

Fig. 3-10. (Left) three-way switch; (right) dimmer switch.

A dimmer switch not only turns lights on and off, it can also vary the amount of current that goes to the light. It works in a similar way to the volume control on your radio or TV.

Connecting Materials

Various connecting devices are needed to attach cable securely to the boxes, and wires to switches, outlets, or each other. The following devices are most commonly used for these purposes and also for attaching cable to the framing.

Solderless Connectors, or Wire Nuts

These are plastic caps with threads inside to grip the ends of the stripped wires (Fig. 3-11). When properly screwed down on the wire ends, these connectors hold the wires tightly together without splicing or soldering. If the wires have been properly stripped, that is, not too much insulation has been removed, no bare copper should be exposed (see pp. 37–38).

Fig. 3-11. Solderless connector or wire nut.

Solder

Used for splicing wires, solder is usually avoided in do-it-yourself situations. If you must do splicing, be sure to use rosin-core solder, never acid-core. Rosin-core and acid-core refer to the flux (an aid to soldering) within the solder itself. Acid flux should never be used for electrical connections because it corrodes the metals of the joints.

Box Connectors

Some boxes have built-in clamps for the cable, and it is easier for the do-it-yourselfer to use this type. If unavailable, however, or if these interior connectors crowd the box too much (see p. 36), you'll need special clamps for each type of cable. Nonmetallic sheathed cable requires a screw down "squeeze" clamp (Fig. 3-12A), and flexible armored cable clamps have a single screw (Fig. 3-12B). Cut off the metal loop strips with tin snips (loops not shown). You will also need fiber or plastic bushings for AC cable connections.

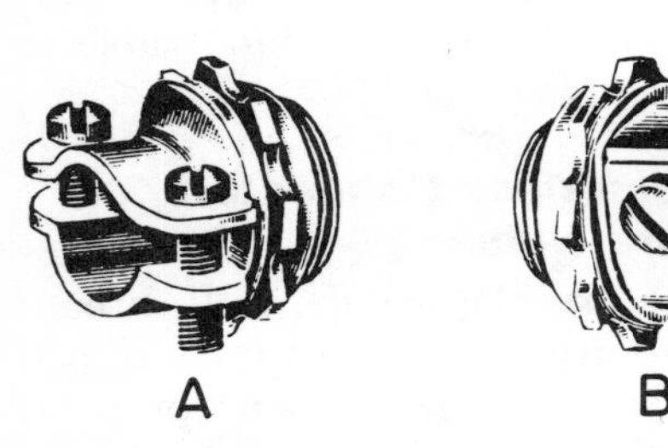

Fig. 3-12. Special connectors are used to secure cable to boxes that do not have internal clamps. (A) A connector for nonmetallic cable, (B) an armored-cable connector.

Black Plastic Tape

Often called electrician's tape and used in many ways. Some electricians routinely tape over the ends of solderless connectors just to make sure that the wire nuts do not loosen. It's not necessary, but a wise precaution if you feel that the connection isn't quite tight enough. Black plastic tape is also good for temporary repair of frayed insulation.

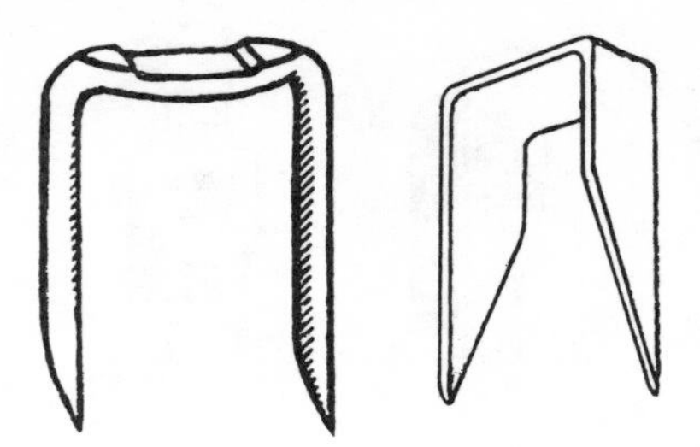

Fig. 3-13. Staples at left are used for attaching nonmetallic (Romex) cable to framing. The type at right is used only with armored cable (AC).

Staples, Straps, and Hangers

Where possible, cable is strung through joists and other framing. When this is not feasible, cable is supported by various hardware, such as special cable staples (Fig. 3-13). On some installations, straps or hangers may be necessary. See the specific installation procedures in Chapter 7.

Conduit Connectors

Highly specialized conductors, discussed in detail on pages 77–79.

Circuit Breakers

As discussed previously, circuit breakers and fuses are the safety valves of your electrical system. Their function is to close down an entire circuit when some abnormality sends a surge of current through the wires. Although quite different in design and method, breakers and fuses serve the same purpose.

Circuit breakers are found in virtually all newer electrical systems, and are safer and more reliable than fuses (Fig. 3-14). A current malfunction will trip the breaker, making the switch flip from on to off. To reset this type of breaker, simply turn the switch back to on.

For high-wattage appliances, such as ranges, two breakers are usually

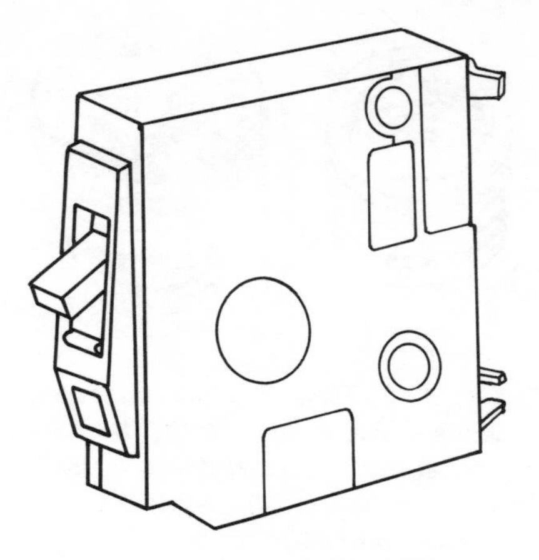

Fig. 3-14. A typical circuit breaker. Circuit breakers come in different sizes and amperage ratings.

ganged by connecting them with a metal bar. The bar is used to reset both breakers at once. The main switch is ordinarily ganged this way, as well.

Sometimes two thin breakers are installed in the same opening in the entrance panel, which means that two circuits run out of there. Only one should trip at a time. Occasionally, buttons are used instead of switches. Press the button to reset this type of breaker.

If you are adding a new circuit, you should know that there are a couple of different types of breakers, neither of which may fit in your entrance panel. Before you buy a circuit breaker, determine the manufacturer of the panel. The "Murray" type is the most common, but some panels will only accept General Electric models.

Fuses

Fuses come in a variety of types and sizes (Fig. 3-15). But, to repeat, you can substitute different types, but never different sizes.

Plug-In Fuses

Common fuses with a threaded "Edison" (light-bulb) base (Fig. 3-16). Modern codes prevent new installations of these fuses because of the possibility of higher amp replacement, as discussed in Chapter 2.

To replace a plug-in fuse, simply remove it and screw in a new one of the same amperage. It is a good idea, when possible, to turn off the main switch before doing this, but make sure there is enough daylight—or use a flashlight—so that you can see what you're doing. You can replace a fuse safely without turning off the main

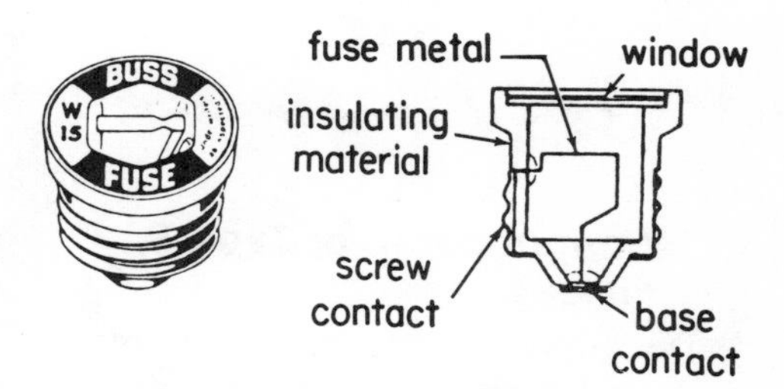

Fig. 3-16. Anatomy of a standard Edison-base fuse.

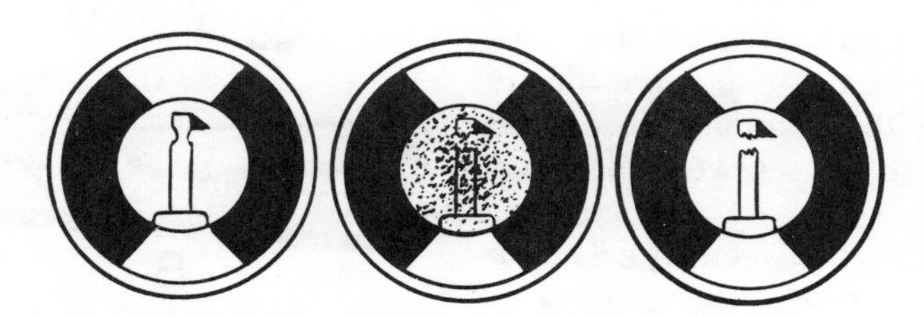

Fig. 3-15. At left, a functioning fuse. The middle fuse shows that a short has occurred, as evidenced by the mottled surface. When there is an overload in the circuit, the copper element in the fuse simply breaks, as shown in illustration at right.

switch if you're careful. Keep your body away from all water or other potential ground. Stand on a dry board to be double-safe.

Time-Delay Fuses

Similar to regular plug-type fuses, except that a springlike metal strip inside allows the fuse to accept a transient overload, such as occurs when starting a large motor (Fig. 3-17). This is not as dangerous as it sounds, since the circuit can handle a brief temporary surge like this. After the motor gets going, it takes a lot less power and poses no danger. A short in the circuit will blow a time-delay fuse the same as any other fuse, and any overload more than momentary will also blow the fuse.

Fig. 3-17. Time-delay fuse.

Nontamperable, Type S, Fuses

Designed to prevent that often-repeated primary evil—replacement of a smaller rated fuse with a larger one (Fig. 3-18). There are two parts to such a fuse: the fuse itself, which is replaceable and much like a regular fuse, and an adapter. The adapter is screwed into the standard fuse box, then the fuse is screwed into the adapter.

The difference between type S and a standard fuse is that the adapter is rated in amperes the same way the fuse is. If you remove a 15-amp fuse from a 15-amp adapter, it cannot be replaced with any other size fuse. Once the adapter is installed, it becomes a

Fig. 3-18. Type S nontamperable fuses come in two parts. The section at left is screwed into a regular fuse socket and cannot be removed. It accepts only one size of fuse like the one at right (also time-delay here).

permanent part of the fuse box, and cannot be removed.

When replacing this type of fuse, make sure that the new one is turned in all the way. The spring under the shoulder of the fuse will not make complete contact unless it is pressed very tightly against the adapter.

Another type of fuse found in residential fuse boxes is the long and cylindrical cartridge fuse, which is similar to, but larger than, those used in automobiles (Fig. 3-19). In homes, these are almost always high-amperage types, used mostly for the main switch.

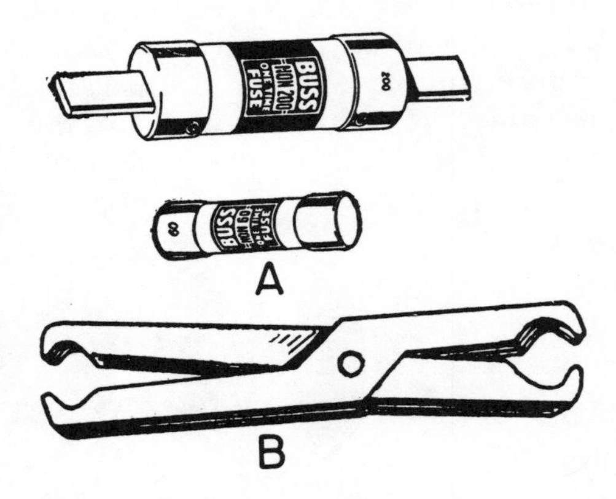

Fig. 3-19. (A) High-amperage cartridge fuses like these shown are dangerous to remove. (B) If you must remove a cartridge fuse, use a fuse puller similar to this one.

Don't tamper with these; call your electrician or power company. If you *must* remove them, use a special cartridge-fuse puller (Fig. 3-19B).

Screw-in Breaker

This is a hybrid, a circuit breaker screwed into a fuse base (Fig. 3-20). There are generally button-type breakers, so that you simply press the reset button when the circuit goes out of commission. The usual reason for installing such a breaker is to prevent frequently blown fuses caused by nondangerous temporary overloads. A workshop circuit is ideally suited for this type of breaker-fuse, where large power tools cause brief, nondangerous surges when started. Perhaps, however, this problem would be better solved by a time-delay fuse.

Fig. 3-20. This looks like a fuse, but it is actually a breaker screwed into a fuse socket. It is useful where a temporary overload is common.

Basic Techniques

Just as carpentry demands such basic skills as knowing how to drive in a nail, electrical work involves a few basic skills that should be mastered before trying more advanced work. The first time we do them, we are usually awkward and slow. But as we progress, they become second nature. Here's how to do them.

Stripping Wire

No matter what type of cable you use, the residential wiring inside is stripped the same way. Stripping means removing the plastic insulation from the ends of the wires so that they can make contact with the electrical device you are installing.

One of the important facts you must remember is that the wires themselves should not be nicked or chipped. Nicked or chipped wires are susceptible to breaking, and a loose wire can cause a short-circuit. Also the current capacity of the wire is reduced in the vicinity of the chip. Damaged wiring can create an electrical hazard.

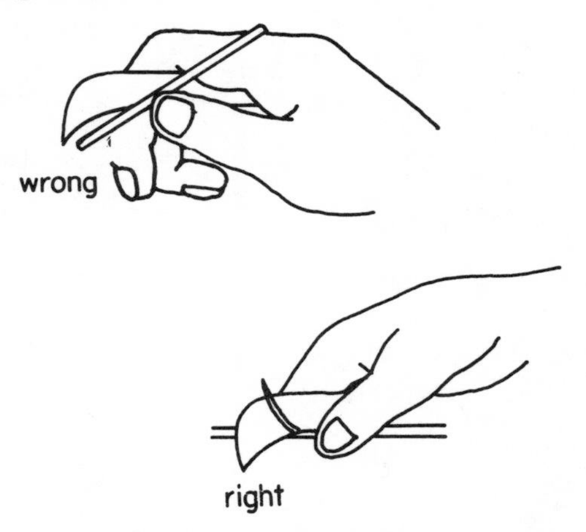

Fig. 4-1. If you don't have a wire stripper or combination tool, a knife can be used to strip wire. Always strip toward you with blade down as shown, and be very careful not to gouge or nick the copper.

Tools

Use a wire stripper or an electrician's tool for this job. A knife *can* be used, but only if you do it carefully. With the blade down, cut from the ends of the wire toward you, as shown in Figure 4-1. The chances of damaging the wiring are much greater with a knife than a commercial stripper (as is the possibility of cutting your fingers).

Wire strippers can be positioned so that they will cut only through the insulation and not the wire itself. They are easier, faster, and safer than a knife. A simple wire stripper shown in Fig. 3-1 is inexpensive and well worth the cost.

Wire-stripping tools are set in different ways so that they will not cut the wire. Each way involves a wire size marked on the tool. If you are strip-

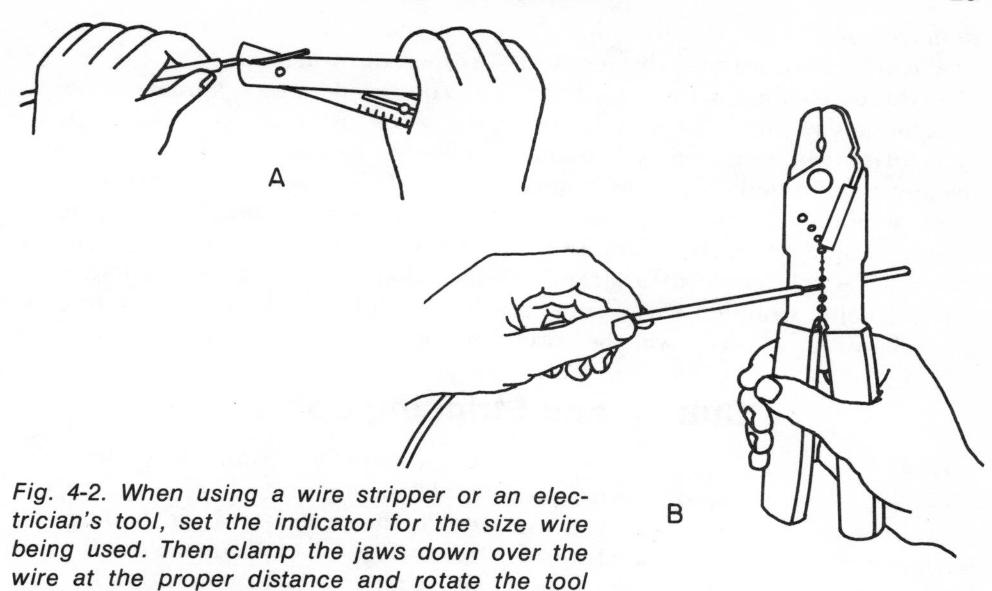

Fig. 4-2. When using a wire stripper or an electrician's tool, set the indicator for the size wire being used. Then clamp the jaws down over the wire at the proper distance and rotate the tool around the wire. Pull off the insulation as shown.

ping #14 wire, for example, look for #14 on an all-purpose tool and insert the wire into the hole next to that number. The wire stripper shown in Fig. 3-1 has settings below the jaws. Place the metal cam, set screw, "arrow" or similar device at #14, and insert the wire into the jaws. The jaws will close only far enough to bite into the insulation, leaving the metal intact.

Although almost any pliers can be used to bend wire, long-nosed pliers do the job most efficiently because it is easier to form the short radius around the thinner ends of this type. The important thing to remember is to bend the wire in the same direction as the screw—clockwise, or so that the open part of the loop is at your right (Fig. 4-3). If you bend the wire this way, tightening the screw will pull the wire snugly around it. A loop facing the other way tends to pull away from the screw. A tight connection prevents loose wires and a possible short circuit.

Technique

With any type of stripper, you will ordinarily have to turn the stripper around on the wiring, then pull forward to remove the insulation.

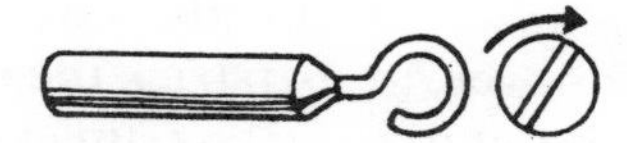

Fig. 4-3. Bend wire to fit over terminal screw, making sure to fit the bent wire over the screw so that tightening the screw does not loosen the wire.

The length of insulation to be removed depends on the device. Most devices, such as receptacles (outlets) and switches, have terminal screws.

Fig. 4-4. Strip enough wire (about ½ to ¾ inch) so that it will extend almost all the way around the screw.

Remove enough insulation so that the wire can be bent around the screw at least three-quarters of the way. Push-in terminals require that only ½ inch of insulation be removed. When connecting wires together with wire nuts, remove about ¾ inch for two solid wires, and ½ inch if you're attaching the solid wire to a stranded wire. Usually, stranded wires, such as in a lighting fixture, will be pre-stripped. If not, remove about ¾ inch of insulation.

You should always loosen the terminal screw as far as you can (without releasing it from the switch or receptacle). That way, the wire can be bent around the body of the screw, instead of trying to slip it over the head. A loop that is wide enough to go over the head will be too large for a tight connection.

Cutting and Stripping Cable

Tools

Cutting and stripping techniques vary with the type of cable you are using. Nonmetallic cable, used by most homeowners, is cut off the roll with a wire cutter or tin snips. The outer sheathing can be stripped before or after it is inserted into the box. It is easier, however, to strip back the cable before it is put into the box, unless you use a cable ripper (see p. 31). The spiral metallic covering of armored cable must be stripped before it enters the box. Then use a hacksaw to cut off AC cable.

Technique

Type NM

Use a utility or razor knife to cut down the center of this cable. Start the cut 7–8 inches from the end of the cable, being careful to keep the cut down the center to avoid cutting the wires (Fig. 4-5A). You may still find an older type of NM cable with a built-in "rip cord" to accomplish the same thing. To avoid damage to the insulation, carefully bend back the outside sheathing and paper wrapping and cut these off with a utility knife or tin snips (Fig. 4-5B).

If you use a special cable-stripping tool, insert the wire into the box first. Then reach inside the box with the stripper to where the cable enters, clamp down on the cable, and pull it back toward you (Fig. 4-6A). The cable sheathing should slice cleanly. Use your razor knife to finish the job, cutting off the sheathing and separation materials inside (Fig. 4-6B).

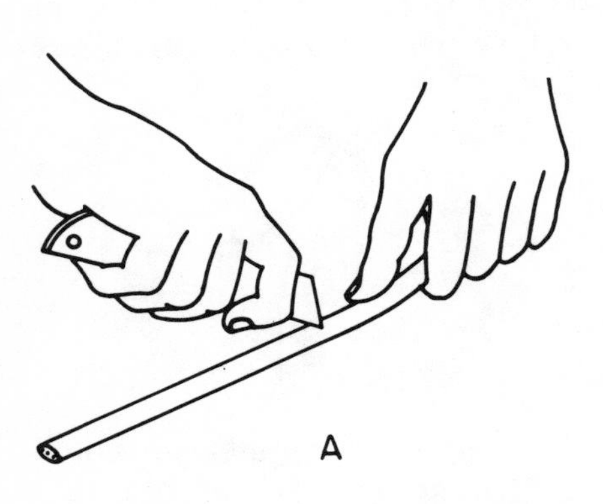

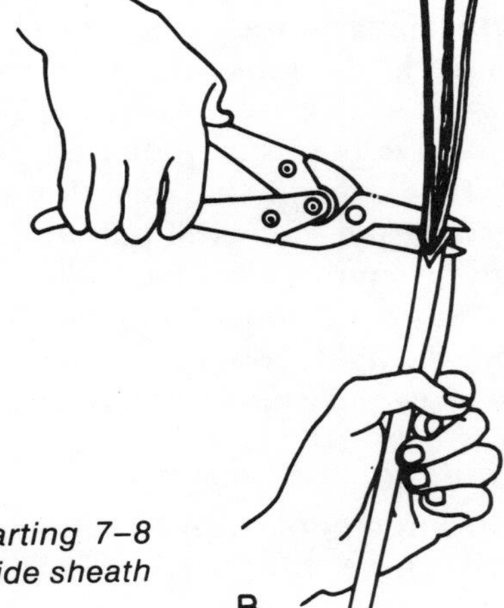

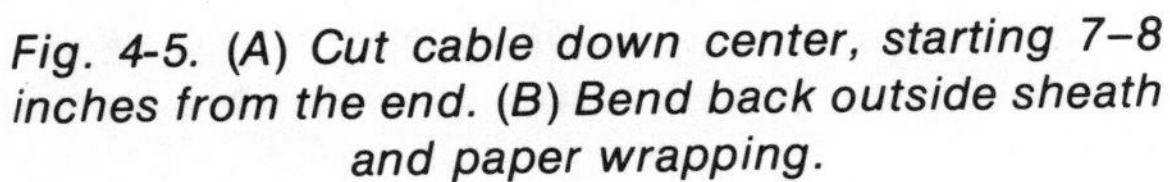

Fig. 4-5. (A) Cut cable down center, starting 7–8 inches from the end. (B) Bend back outside sheath and paper wrapping.

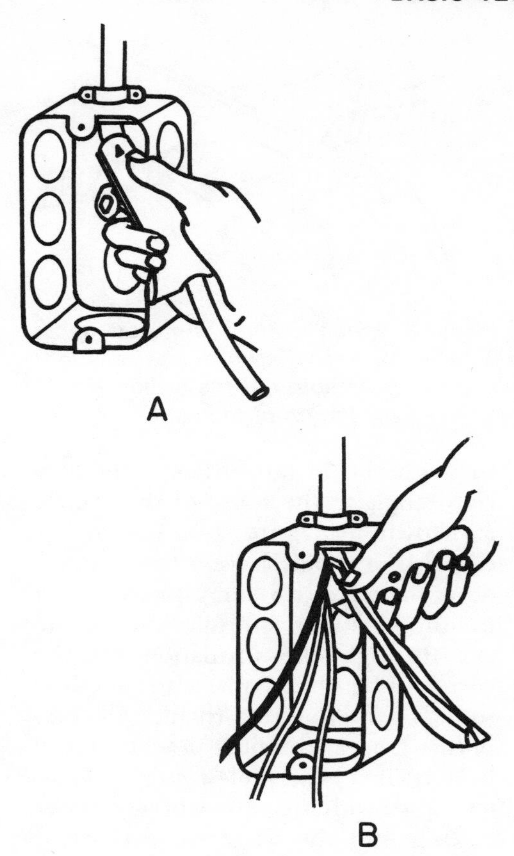

Fig. 4-6. *Novices are usually better off stripping cable before inserting it into the box. (A) You can, however, attach the cable to the box first and use a cable ripper to make the center cut as shown. (B) After the cable ripper has done its work, finish removing the plastic sheathing and paper.*

Type UF

Since underground nonmetallic cable has a tough, solid plastic sheath, Type UF cable is more difficult to cut and strip than Type NM. The techniques, however, are the same. You will have to cut firmly and carefully with your knife, however, to free the wire without damaging the insulation. Be sure, in this case, to have the cable stripped and ready to work with before putting it into the box. It is very difficult to do this afterward.

With Type UF cable, try to have a hard surface to work on, so that you apply maximum pressure with your knife. An old board is a good cutting surface, because you can press hard against it, without worrying about damage if the knife slips. Be prepared to use a little extra patience when freeing the wires from this type of cable. (See Figure 10-6).

Type AC

Flexible armored cable (BX) is protected by a galvanized steel jacket, wound in spiral fashion so that it can be bent around corners. Inside the steel jacket, each wire (except for the ground) is wrapped with strong paper. There are two accepted ways to prepare armored cable for connections to the box, after cutting it off with a hacksaw.

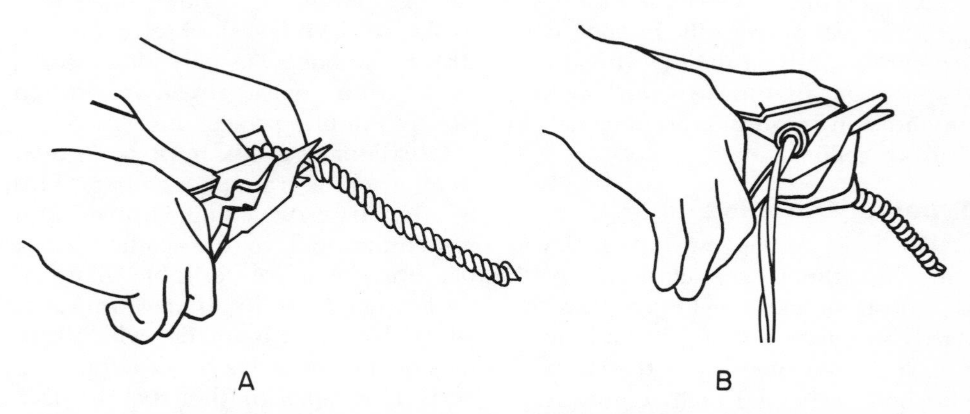

Fig. 4-7. Drawings show the steps in removing the metal armor from AC cable.

The best method is to grasp the cable at the point where it will be inserted into the box, and bend it sharply until you feel the armor buckle. One of the spiral turns should bulge outward. Grip the cable on both sides of the buckling point, and twist against the direction of the spirals until the spiral bulge moves out enough so that you can slip the jaw of a metal snips underneath (Fig. 4-7).

Use the metal snips to cut through the armor at that point, cutting until you can slip off the outer end of the armor, taking care not to damage the enclosed wires. Trim off any jagged edges at the edge of the steel armor. If the armor has been bent or distorted, use the large jaw recesses of the pliers to reshape it. Shape the final turn of the spiral to conform to those behind it, using the tips of the metal snips. The cable is now ready for bushing (Fig. 4-7). Remove paper wrap, if used and slip the fiber bushing over the wires and push it back to the edge of the armor (Fig. 4-7).

A hacksaw can also be used to cut through the armor. Always cut at a right angle to the spiral of the armor, as shown in Figure 4-8, not to the cable. If you only have a few connections to make, this may prove easier for the beginner, but take care not to cut through the insulation on the wires. The bend-and-snip method sounds a little more difficult and complicated, but with a little practice, you'll be able to cut the cable faster and easier, and without any worries about cutting into the wires themselves or their insulation.

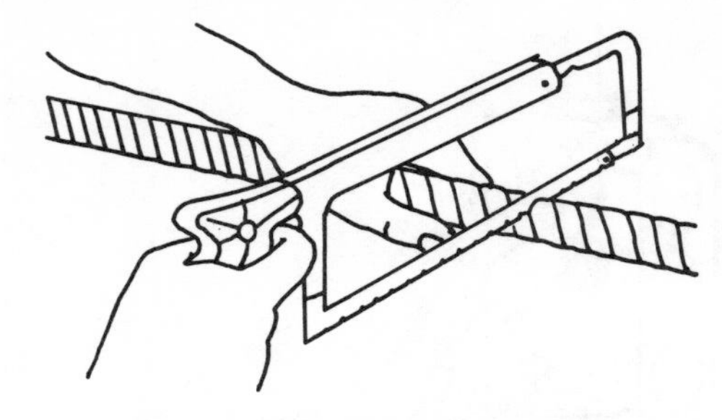

Fig. 4-8. When cutting armored cable with a hacksaw, always saw at right angles to the spiral of the armor, not to the length of the cable.

Mounting Boxes

There are a great many different types of electrical boxes, but basically only two ways of attaching them to walls and ceilings. The easiest method, used almost universally in new construction, is to nail the box directly to the framing of the new work before any finishing materials are attached to walls or ceilings.

Types of Devices

The most convenient boxes, if you can find them, are equipped with mounting brackets welded to the box itself. Simply nail through the bracket into the front or sides of the studs or joist bottoms with 1-inch roofing nails (Fig. 4-9A). Other boxes are nailed with 8d (8-penny) nails into the sides of the studs through projections in the top or bottom, or through holes predrilled in the boxes themselves (Fig. 4-9B). Some boxes, usually plastic ones, come with nails already attached through in-line projections.

Occasionally a box must be located away from the framing members. This is often true of ceiling fixtures, and sometimes wall fixtures, when exact placement is more desirable than it is for a switch or outlet. In new work, use adjustable bar hangers, which are nailed into the studs or joists on each end. The box can then be slid and

Fig. 4-9. (A) It is easy to nail boxes with attached brackets. Simply nail through the brackets into the studs with 1-inch roofing nails. The box edge should be lined up flush with the exterior of the intended finishing materials. (B) For boxes without brackets, nail 8d (8 penny) nails through the predrilled holes in the sides of the box.

locked in place at the optimum location. For ganged boxes, you will also need wood or metal mounting bars.

Technique

When you are working with existing walls or ceilings, box-mounting, like everything else in old work, is a little more difficult. When the proper location of the box is determined, a hole is cut into the wallboard or paneling to accept the new box. Make a paper or cardboard template of the box by laying it face down and tracing around it. (Some box manufacturers supply a template with the box.) Trace around the template onto the wall to mark the rough opening. If only one or two boxes are involved, it may be simpler just to hold the box itself to the wall and trace around it.

If the walls consist of gypsum wallboard or paneling, drill holes about ½-inch in diameter at the corners of the box opening and cut out the opening with a keyhole saw. When the walls are made of real plaster, chisel away some of the plaster near the center of the box first (Fig. 4-10). If there is metal or gypsum lath behind the plaster, proceed as above for regular walls, but use a fine-toothed blade, such as a hacksaw blade, to avoid damaging the plaster. In homes built prior to World War II, you will probably find wooden lath behind the plaster. If so, chip away a little more until you expose a couple of pieces of wood lath. Then adjust the box location, if necessary, so that the top and bottom of the boxes will fall in the middle of the lath strips. The lath strips are about 1½ inches wide. Cut out the opening as above with a fine-toothed blade. Then chip away about ⅜-inch more plaster above and below the opening to allow direct mounting of the box to the wooden lath with #5 wood screws.

Special Mounting Devices

For all other walls, special mounting devices will be needed. There are several types, many of which are attached to the boxes themselves. Some have

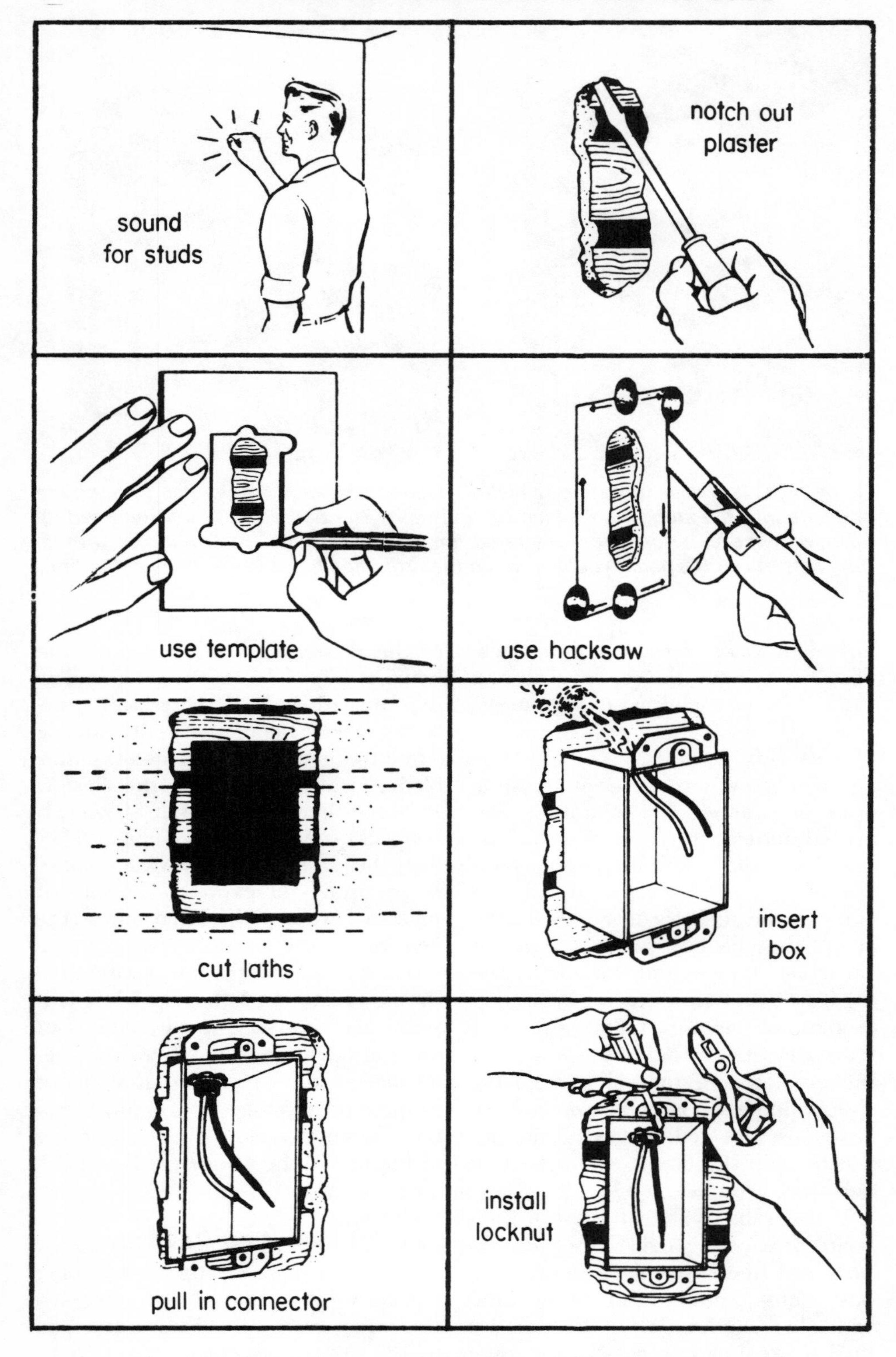

Fig. 4-10. Steps taken to attach boxes to lath and plaster walls.

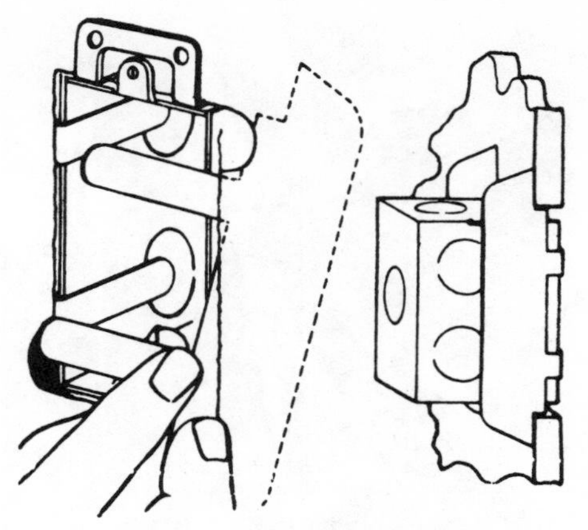

Fig. 4-11 "Madison clips" are used to attach boxes to existing gypsum wallboard. Insert clips between box and wall on both sides as shown at left, and bend the tongues back inside of the box. At right, the installed box from inside the wall.

clamplike devices that hug the back of the wallboard when the attached screws are turned. Ask your hardware or electrical dealer for this or a similar type of box. Boxes without mounting devices can be attached to wallboard or thin paneling with "Madison clips," which are slipped between the box and wall on both sides, then bent back over the insides of the boxes as shown in Figure 4-11. (The longer length at top and bottom keeps the clips from falling out.) On thick paneling, boxes can be screwed directly to the wood.

Ceiling boxes in old work should be mounted from above where possible, as in an attic, using adjustable bar hangers. (See Chapters 7 and 8 for details)

Connecting Cable to Boxes

Professional electricians connect the cable to the box before stripping it, but you will probably find it easier to at least remove the exterior sheathing first. In awkward places, it may even be wise for you to remove the insulation from the wires before attaching to the box.

No matter when you do the stripping, leave enough cable inside the box to give yourself room to make the connections afterward. Although 8 inches is generally recommended, this is a lot of cable to stuff back inside once you've finished stripping, particularly in the middle of a circuit run, where wires run both in and out of the box. Local codes may even specify 8 inches, but you can usually get by easily enough with only 6 or 7 inches.

The actual connections between cable and box depend on the type of cable and the type of box, and you may find some variations on the methods described below. Your dealer can tell you what technique to use if the ones below don't apply. In general, however, these are the techniques you should be familiar with:

Technique

Nonmetallic Cable (Romex)

Attach both Types NM and UF to interior boxes by clamping the wires. When attaching cable to metal boxes, remove the knockouts on the box where the cable will enter by twisting them out with a screwdriver or by tapping with a hammer. If the box has built-in clamps, loosen the setscrew that holds down the clamp, insert the cable to the proper length, and turn the screw in again until the cable is held tightly (Fig. 4-12A). Most clamps go over two knockout holes, so insert both incoming and outgoing cables beforehand if you use both of these knockouts (4-12B).

Fig. 4-12. (Left) With built-in cable clamps, loosen the clamp and insert the cable. (Right) After the cable is in place, tighten clamps until secure. If two cables enter the box through the same clamp, be sure both are in place before tightening.

Connectors

For boxes without built-in clamps, purchase either metal or plastic clamp-type connectors. Metal connectors, which are the more common of the two, come in two parts. Remove the locknut, then slip the clamp part over the cable before inserting into the box (Fig. 4-13A). Turn down the set screws to hold the cable in place, and insert the connector into the box knockout. Hand-tighten the locknut over the clamp portion, then secure firmly by tapping a screwdriver blade against the "points" of the locknut (Fig. 4-13B). With plastic connectors, firmly

Fig. 4-13. (Left) Slip the clamp section of an exterior connector over cable before putting it into box. Tighten the set screws to hold it in place on the cable. (Right) Tighten the locknut over the connector to keep it in the knockout. Place screwdriver against locknut and tap screwdriver handle with a hammer to thoroughly tighten.

push through the knockout hole, then insert the plastic wedge into the slot provided.

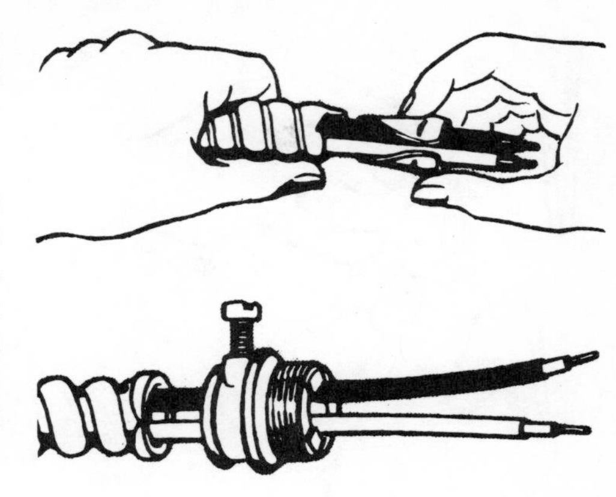

Grounding

Since nonmetallic boxes do not present a danger of grounding if a cable is loosened, connectors are not necessary (but check local codes) as long as the cable is stapled to the framing within 8 inches of the box. If that is not possible, use a nonmetallic box with built-in clamps.

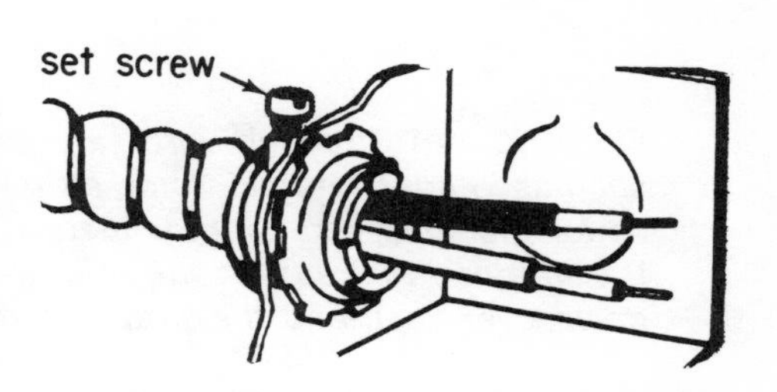

Armored Cable (AC)

Armored cable requires the use of metal boxes for proper grounding. Buy boxes with special built-in AC connectors, or use one of the several types of connectors bought separately. All of these connecting devices work like the NM connectors discussed above, except that a set screw holds the cable instead of a clamp. Don't forget to use a plastic or fiber bushing at the end of the stripped armor (Fig. 4-14 top).

Fig. 4-14. AC cable connectors are similar to NM connectors, but only one set screw is needed. Be sure to install the fiber bushing (top) before installing the connector.

Joining Wires

When two wires are to be joined together rather than connected to a device, the most convenient way of joining the wires is by using a "wire nut" or solderless connector. These devices come in several sizes, so be sure to ask your supplier for the right size. In general, the smaller ones are for light, stranded wires such as those found in lighting fixtures. Medium-sized wire nuts are the most commonly available, and these are the right type for the size wires you will no doubt be using (#14 and #12).

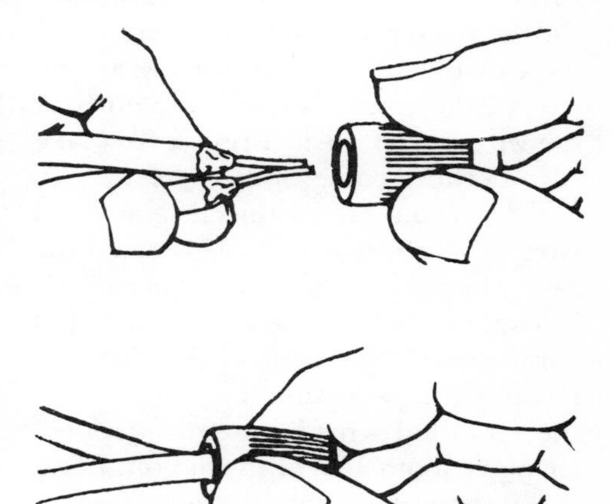

Fig. 4-15. To attach two solid wires together with a "wire nut" or solderless connector, hold the wires together as at top then "screw" on the connector by turning it clockwise over the wires.

Technique

Wire Nut

To attach solid wires together with wire nuts, strip off ½ to ¾ inches of the insulation, making sure that the

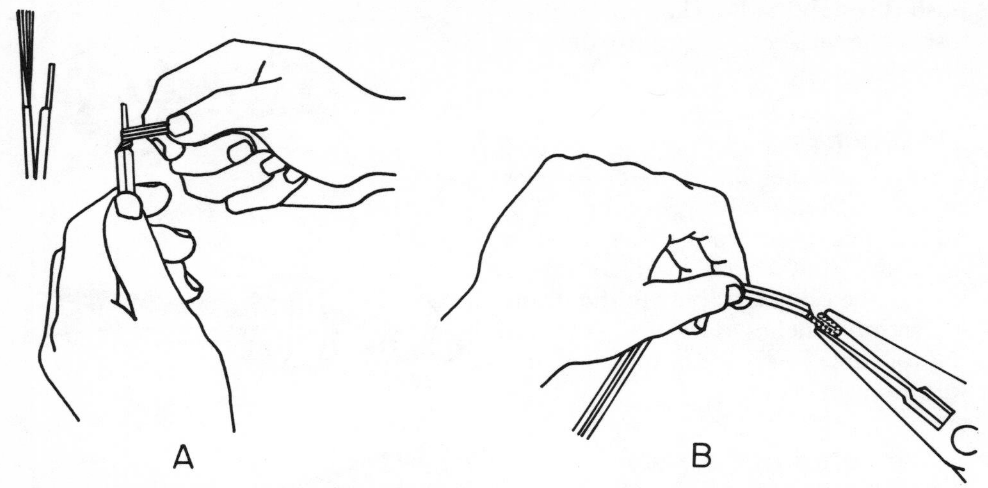

Fig. 4-16. (A) Stranded wires are usually prestripped for common use, such as in light fixtures. If not, strip back about twice as much insulation as for the solid wire (approximately 1½ inches) as shown in inset. Wrap the stranded wire around the solid wire. (B) Bend the end of the solid wire down over itself and the wrapped stranded wire and screw on the wire nut.

stripped sections are even. Hold the wires together and "screw" the wire nut down over the ends of the two wires by turning it clockwise (Fig. 4-15). Give the connector a gentle tug to make sure that the wires are held firmly. (Don't yank hard. They will surely come apart.) If any bare wire shows, wrap the area with plastic tape. As a matter of fact, you can wrap the end of the connector and the ends of the wires with tape in any case, to make sure nothing comes loose.

When you are connecting a solid wire to a stranded wire, as is often the case when you hook up a wall or ceiling lighting fixture, wrap the stranded wire around the solid wire before attaching the wire nut (Fig. 4-16A). If the stranded wire is not prestripped, strip off about 1½" of insulation. Bend down the end of the solid wire over itself as shown in Figure 4-16B. If three or more wires are to be attached together, you'll need an extra-large wire nut. It is difficult to get more than two wires to hold together properly with a wire nut, however,

Connectors for Three or More Wires

I like to use special connectors for three wires or more. These come in two parts. One is a metal ring which slips over the wires and is tightened with a set screw. The wire nut is then screwed onto the threaded outside of the metal ring. Another device is a compression ring, which is placed over the wires and squeezed with a crimping tool (or the crimping section of an all-purpose electrician's tool). A special insulating cap is then pushed over the compression ring and the ends of the wires.

Splicing and Soldering

Solderless Connectors

For practically all your wiring needs, solderless connectors will do a safe and proper job. There should be little, if any, need for splicing wires together. Indeed, in most cases, splices are

Fig. 4-17. When soldering wires, bare each one for about 3 inches, tapering the insulation back about 20 degrees. Starting about 1 inch beyond the ends of the insulation, twist each wire around the other six to eight times.

against the Code. You can never, for example, make a splice outside of a box. If cable connections must be made where there is not a receptacle, switch, fixture, or other natural breaking point, a junction box must be used.

There may be an occasion, however, where you will have to splice some wires. This could be true where there is some extra strain on the cable, or the cable cannot be securely attached with connectors. If and when splicing is necessary, it must be done correctly. A good splice should be as strong as the wire itself. When it is not, there could be trouble.

Splicing

To make a proper splice, bare each wire for about 3 inches, tapering the insulation back about 20 degrees as shown in Figure 4-17A. All wires should be clean and shiny. Cross the wires (black to black and white to white) about 1 inch beyond the insulation on each end, then twist the wires six to eight times tightly around each other (Fig. 4-17B).

Place a soldering gun or iron on the wires, with the end of the rosin-core solder nearby, and heat until the wires are hot enough to melt the solder (Fig. 4-18). Don't apply so much heat that it melts the insulation. The solder is soft, and doesn't require boiling heat. Slide the solder along the wires so that it flows into every crevice, completely coating each wire. Let the soldered joint cool naturally, without disturbing the molten solder; otherwise a crystallized joint will result. Crystallized joints are mechanically weak and electrically unreliable. You can recognize

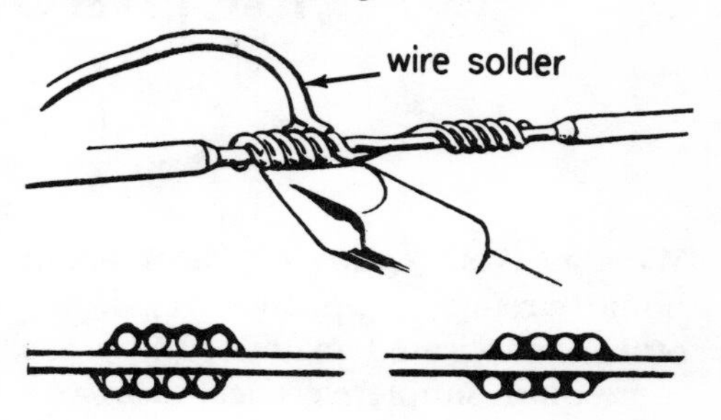

Fig. 4-18. Apply the soldering gun or rod to the wires themselves, not the solder. Heat until the solder melts and starts to flow onto the hot wires. When properly soldered, the splice looks like the one at bottom right, not superficial as at bottom left, which was applied to a cold wire.

such a joint by its dull, often rough appearance. When the solder cools, cover the splice from one end to the other with plastic electrical tape, stretching the tape tightly (Fig. 4-19). Keep taping back and forth over the splice until it is about as thick as the insulated wire, or perhaps a little thicker. Tape about ½-inch back onto the insulation at both ends.

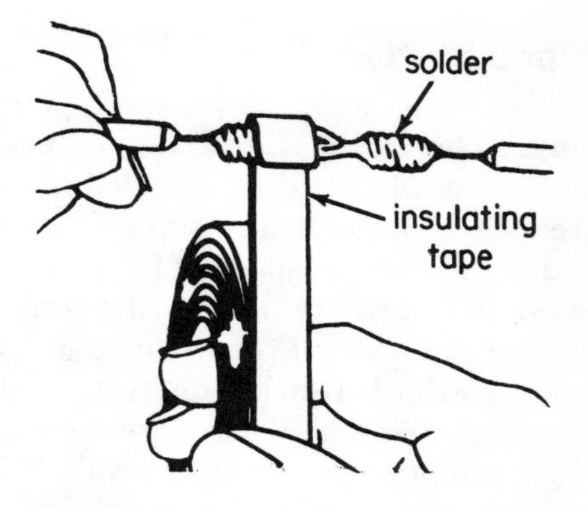

Fig. 4-19. With black plastic tape, wrap layers over the splice until it is as thick as the wire insulation or a little thicker.

Simple Repairs and Replacements

Whether you know (or care) about home circuitry, appliance repair, or other complicated matters, there are some quite simple electrical tasks you can do around the house. This chapter covers repairs and replacements that can be attempted by the novice do-it-yourselfer, regardless of his electrical knowledge.

Replacing Plugs

This is a basic task that everyone runs into, even an apartment dweller. It is something the most craven of electrical cowards can do. The problem results from a break in the plug or at the end of the cord in a lamp or appliance. It is common because, no matter how many times we're told not to do it, we all pull on the cord instead of the plug. This loosens the wires in the plug itself or results in cracking of the insulation near the plug, eventually causing a short or a break in the wires themselves.

Types of Plugs

The repair depends on the type of plug. There are two basic types—molded and heavy-duty. Molded plugs are plastic without any screws inside and are found on zipcord (Type SPT). Molded plugs are ordinarily found only on lamps and other low-wattage items which use the thin multistranded wire in zipcord. The wires are molded right into the plug. As a result, the plug itself cannot be fixed. Dire though this may sound, repair is deceptively simple. Every hardware store, and even many drug stores or supermarkets, sell replacement plugs in their electrical departments. (Look for the display where light bulbs and fuses are sold.)

Molded Plugs

There are different types of molded plugs available, all with complete instructions, so always check the directions carefully before beginning work.

Some plugs are sized for specific cord sizes (e.g., #14 or #18) and some require that one of the wires be cut shorter than the other by a specified length.

The two basic types of molded plugs are one-piece and two-piece. With both types, start by cutting off the old plug and any portion of the wire that is cracked or damaged. For the one-piece type, raise the plastic lever on top, then stick the cut-off end of the wires into the side of the plug through the slot provided. Push down on the lever to engage the prongs into the two wires, and you're done.

The two-piece types are a little more complex. With this type, insert the

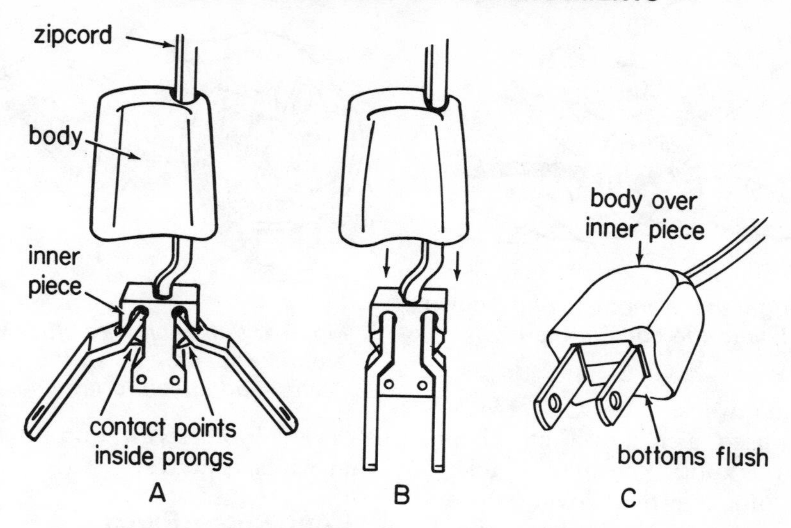

Fig. 5-1. All types of replacement plugs for zipcord make contact by pressing contact points into the wires as shown at left. With two-piece types, slide the outer section up the cord, insert the end into the inner section and press prongs in as shown at center. Squeeze the outer section over inner portion to hold in points (right).

cord through the hole of the "bell-shaped" outside piece. Slip the cord end into the slot of the rectangular inner piece with the prongs. Now slide the outer piece over the inner piece, applying enough pressure to push the prongs into the wires. The ends of the prongs should be straight, with the two pieces level at the bottom (Fig. 5-1.).

Heavy-Duty Plugs

Toasters, irons, and other appliances generally have larger wires and heavier duty plugs. These plugs, shown in Figures 5-2 and 5-3, have standard screw terminals for the wires, much like the screws in an outlet receptacle.

If the plug itself is not damaged, it can be reused by removing it, cutting off the damaged end of the cord, and reattaching the plug as described below. Damaged plugs will have to be replaced, but removal and replacement are the same for new or old plugs.

First remove the fiber cover from the prongs, prying off lightly with a knife point or a screwdriver. Loosen the terminal screws to release the wires, then try to pull off the plug.

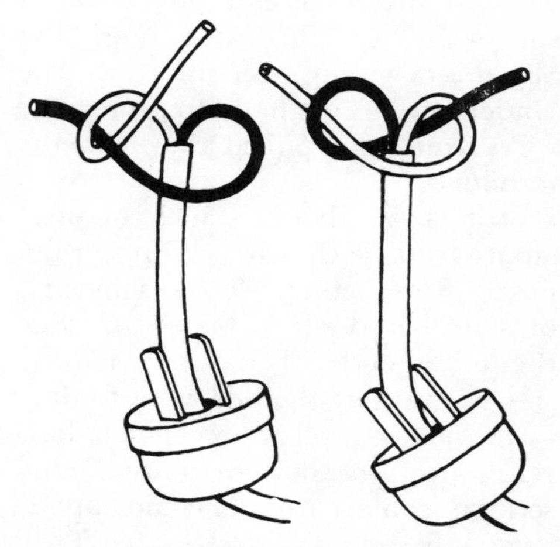

Fig. 5-2. Heavy-duty plugs require an Underwriters knot. With 3 inches of separated wire, loop the black wire through the white one (left), then slip the white wire over the end of the black and back up through the loop in the black (right).

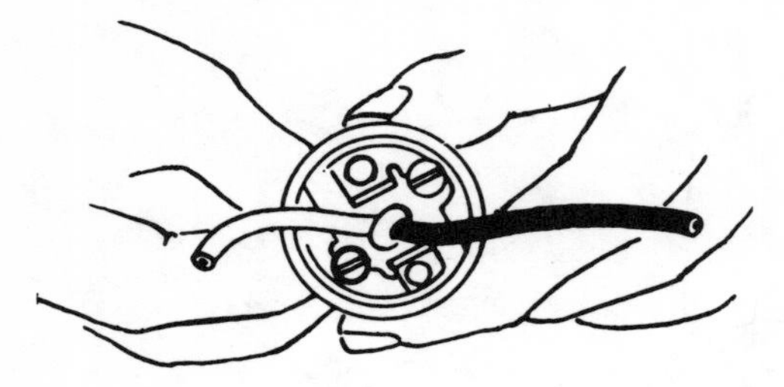

Fig. 5-3. Tighten the knot and pull it down snugly into the body of the plug.

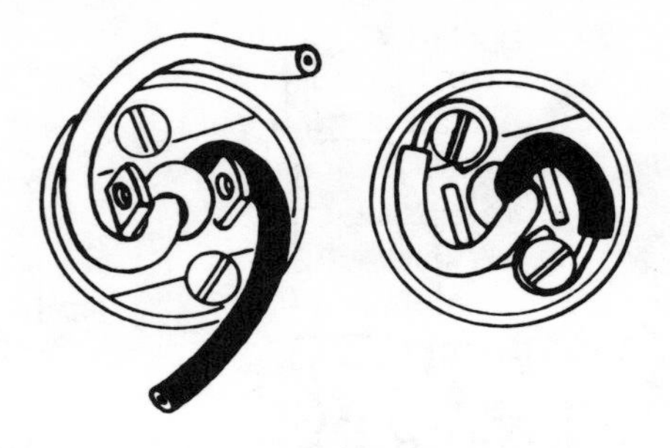

Fig. 5-4. Pull each wire clockwise around the prongs of the plug, strip and attach to terminals.

This won't work if an Underwriters knot was used, as it should have been, so untie the knot. (You may be able to retie the knot later by following the old bends.) To tie an Underwriters knot, first separate about 3 inches of wire. Loop the black wire behind itself and the white wire, then slip the white wire over the end of the black and bring it up through the black loop (Fig. 5-2). Tighten the knot and pull it down snugly into the plug body. (Fig. 5-3.) Slip the new plug over the cord. The Underwriters knot helps to protect the wires from being pulled away from the terminals.

Strip away about ¾ inch of insulation from each wire. With a thin-nosed pliers, bend the loops in the ends of solid bared wires. Make sure that the loose ends of the bared wires stay where they should—next to their own hold-down screw—and don't reach over to shortcircuit the other screw or contact area. This can happen if you bared too much wire. Twist together any stranded wires. Fit the wire inside the plug and place the loops clockwise around the terminal screws—black to brass, white to nickel-colored (Fig. 5-4). Tighten the screws, then replace the fiber insulator over the prongs.

To replace a three-prong adapter plug, tie all three wires together in a tight knot, and pull the cord until the knot is snug against the plug. Loop the green (ground) wire around the green or dark-colored screw, and attach the other wires as above.

Appliance Plugs

Many appliances, such as electric coffee pots, have a separate cord. A heavy-duty wall plug is on one end, and a female appliance plug on the other (Fig. 5-5). Wall plug replacement is as described above. If the appliance plug or the cord at that end is damaged or worn, it too can be replaced.

Most new appliance plugs are riveted together and cannot be reused. They must be removed by cutting the cord and buying a replacement. If your appliance has a plug that is held together by screws or bolts, and the problem is with the cord inside the plug or near the top of the plug, simply cut back the cord and reinstall the old plug. Otherwise, replace the plug.

Use a knife to carefully remove

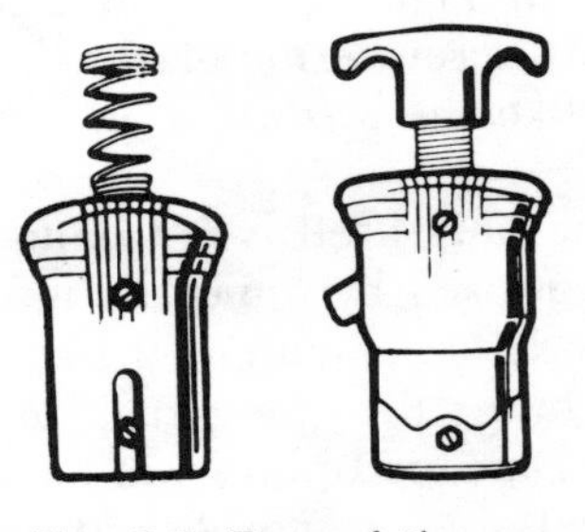

Fig. 5-5. Two of the several types of appliance plugs, the one at left with a spring.

about 2½ inches of outer insulation from the cord end. Strip off about ¾ inch of insulation from each wire, taking care not to damage the wire. Twist each multistranded wire tightly, then heat with a soldering gun or iron, applying a small amount of solder to each. Trimming the wires in this way makes it easier to attach them.

Remove the screws holding the halves of the plug casing together. Insert the cord through the spring guard as shown in Figure 5-6. Bend the wire ends into loops and place them under the terminal screws so that tightening the screws (clockwise) will close the loops. Tighten the screws. Put the spring guard, if any, back into half of the casing. Rescrew or bolt the casing halves together.

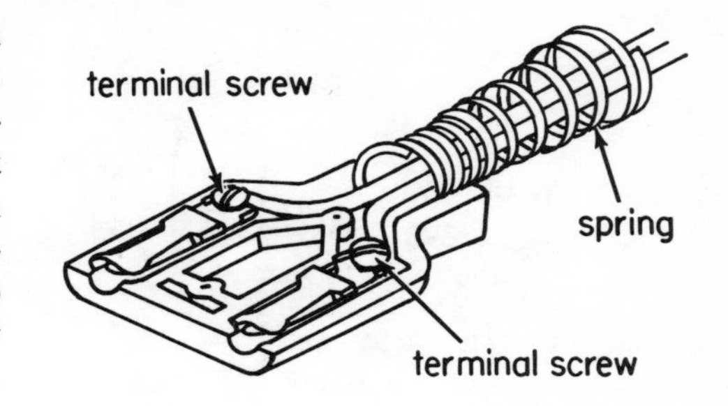

Fig. 5-6. The appliance cord goes through the spring and is attached to the terminals, which are exposed by unscrewing and separating the two halves of the appliance plug, shown here with top half removed.

Replacing Cords

If a cord is damaged so far from the end that it will be shortened too much by repair, throw it away (the cord, not the appliance). Frayed insulation is another cause for consignment to the scrap heap. Sometimes an extension cord will be damaged because too many appliances are run off it. The wires become too hot and the insulation melts. Get rid of it.

Cord damage is not always obvious. The insulation may look okay, but wires can be broken inside. One probable sign of this condition is that the lamp or appliance goes on and off as you move the cord, making intermittent contact between broken wires. The cord probably needs replacement.

When replacing any cord, always buy the same type and size as the original. Thin lamp cord cannot carry the high-wattages of such heat-producing appliances as toasters and irons. These require "heater" cord, (Type HPD) wrapped with asbestos layers inside cotton or rayon braid. Some may use Type HPN, which is like a heavy "zip-cord." If you are in doubt, take along the old cord when buying a new one at the hardware or electrical supply store.

Repairing Incandescent Lamps

Incandescent lamps (using standard filament bulbs, as opposed to fluorescent tubes) come in innumerable varieties of styles and shapes, but they are all alike electrically. Fanciness and trends notwithstanding, all a lamp really needs is a socket to hold the bulb, a switch to turn it on and off, a cord to the power, and a plug to hook into an outlet. Figure 5-7 shows how the business end of a typical lamp is put together.

Assuming that there is nothing wrong with the entrance panel or outlet when a lamp doesn't work, the first thing to check is the bulb. Obvious though it might sound, this step is sometimes forgotten. The best way to do this, of course, is to put in a new bulb. But suppose you're fresh out of

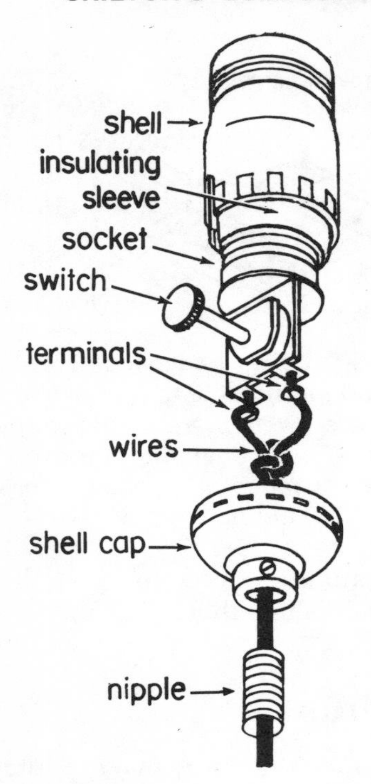

Fig. 5-7. The business end of a lamp.

bulbs? How do you know if it's the bulb or something in the lamp? One good old test—shake the bulb next to your ear and listen for a piece of filament rattling around. Or unscrew a bulb from another lamp and test it. If it lights, at least you know the problem. Never, by the way, leave a bulbless lamp plugged in. It's too easy for a child (even an adult) to stick his finger inside the socket and get a sizable jolt.

Cord Replacement

Let's suppose, however, that the trouble is not in the outlet or bulb. Next check the plug and the cord, as explained above. If either is faulty, fix it, as described. You may run into a problem, however, if there is a break in the cord that is not near the plug. Replacing an entire lamp cord can be difficult if the lamp is complicated. In such cases, or if there are a lot of twists and turns inside the lamp base, it is a good idea to use the old cord as a "fish tape" (more or less).

To do this, remove the socket shell and cardboard insulating sleeve as described below. Loosen the terminal screws and set screw in the socket cap and take off the old wires. Untie the Underwriters knot. Strip about ¾ inch insulation off the ends of the new wires and twist the old and new wires together. Warp the bared wires with a few twists of tape for added strength and to prevent their catching on something inside the lamp.

When this is done, start pulling on the old cord from the bottom of the lamp. Go slowly, at least at first, and avoid jerking. As you pull out the old cord, the new one will be threaded properly through the lamp. When the new cord appears at the bottom, re-

Fig. 5-8. A harp-type lamp, with built-in switch, before dismantling.

move the old one and keep pulling on the new one until the other end is where you want it. Leave about 2½ inches so that you can tie an Underwriters knot. Then strip and attach that end to the terminals. Replace the insulating sleeve and socket shell.

Switch and Socket Replacement

If you know that the bulb, plug, and cord are okay, then the trouble is most probably in the switch (Fig. 5-8). It could be the socket, of course, but it makes no difference, because they are replaced together. New sockets are commonly available in hardware, electrical supply, or even variety stores. Loosen the old socket shell by pressing with your thumb where it says "Press" and lifting up, using a twisting motion (Fig. 5-9). Pull up the insulating sleeve, and loosen the wires from the socket. Take the old socket with you when buying a new one, to make sure you get the right type. The new one will include the socket cap, at the bottom of the socket, but there is ordinarily no need to replace this part.

As long as you don't have to replace the old socket cap and thereby untie the Underwriters knot in the wires, all

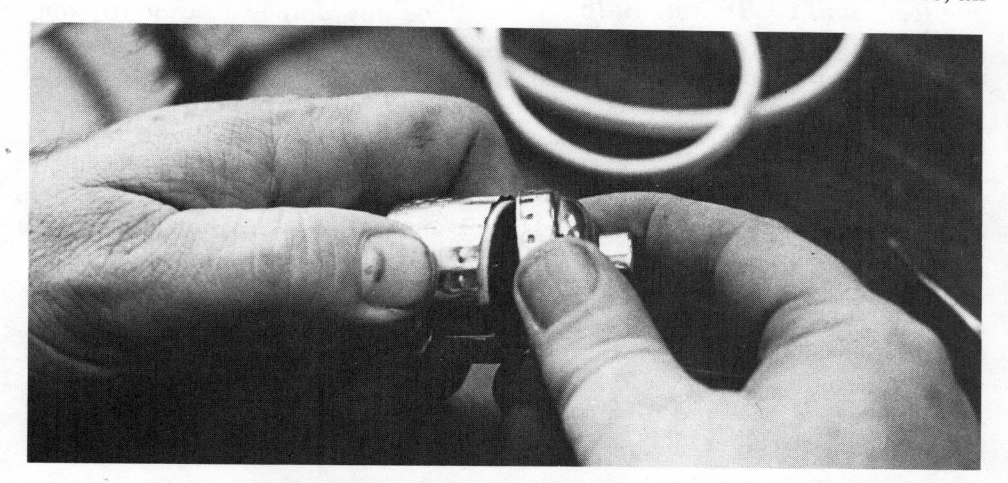

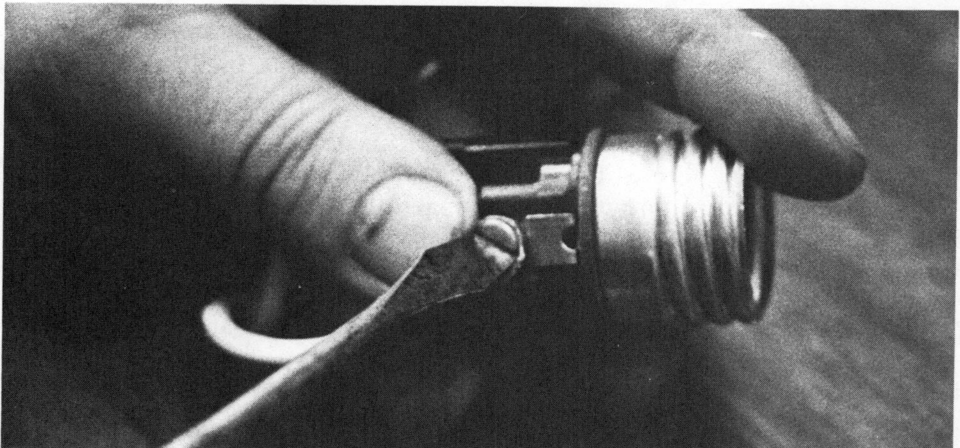

Fig. 5-9. (Top) To remove the socket shell, press down where so marked on the shell and pull and lift from the cap. (Bottom) Bend the wires around the terminals in clockwise direction and tighten the screws.

you have to do is reattach the wires to the terminals of the new socket, replace it in the old cap, put on the cardboard sleeve, and put in the new shell (Fig. 5-9). Line up the corrugated edges of the shell with the cap, press where it says to do so, and push down until you hear the parts click together.

If you must replace the socket cap, loosen the set screw at the bottom, untie the Underwriters knot, and put the new cap where the old one was. Turn in the set screw until it is firm, but not too tight. Retie the knot and proceed as above.

Removal of Broken Bulb Base

Sometimes a light bulb breaks inside a socket and it can be difficult to remove the bulb base from the socket. There are a couple of ways to remove the base. One is to stuff some old newspaper or rags inside the base and press down while turning to dislodge the base. If that doesn't work, stick in a screwdriver, pliers, or something similar with the point pressing against the bottom of the base. Press also against the edge of the base and turn (counterclockwise, of course). Or grab the edge of the broken-off base with the tips of a long-nosed pliers and gently work the shell loose.

And, speaking of "of courses," need I tell you to pull out the plug of the lamp whenever you're doing anything more than routinely changing the bulb? Yes, I'm afraid I must. As a matter of fact, it's not a bad idea to unplug the lamp even while just replacing the bulb. (You usually don't know if it's "off" or "on.") It's a must if you leave the lamp unattended while you search for a new bulb. Also if you're working on a chandelier or other permanently wired fixture, kill the circuit by removing the appropriate fuse or tripping the breaker.

One more thing. On many lamps, you can fix whatever needs fixing without removing the harp. If it gets in the way, however, it can be removed by first pressing the two sides together just above the socket. Many harps have small caps which slide down and cover the places where the harp snaps in; these caps must be pushed up out of the way before you can squeeze the harp sides together. This should separate the top part of the harp from the base for easier access.

Repairing Fluorescent Lights

The energy-saving features, and other advantages, of fluorescent lighting are discussed in Chapter 12, so we won't dwell on them here. Suffice it to say that, it pays to care for and repair fluorescents. Do not use them in unheated garages or basements in cold climates. They don't work well below 50° F.

Because of their generally unstylish look, fluorescents are usually found in dropped kitchen ceilings, under counter-tops, in "student" lamps, and in the bathroom.

The heart of the fluorescent lamp is the "ballast," which is really a small transformer. A fluorescent tube contains no filament, so the ballast is necessary to set up voltage inside the chemically active tube, as described on pages 48–49. Older fluorescent lamps were also equipped with a starter, which gave an extra boost to the ballast in the start-up process. Newer lights have starter mechanisms built right into the ballast.

Unlike incandescent lamps, which may be complicated to repair because of the wide differences in styling, fluorescent lamps tend to be similar and more utilitarian. The most common 2-foot and 4-foot models have

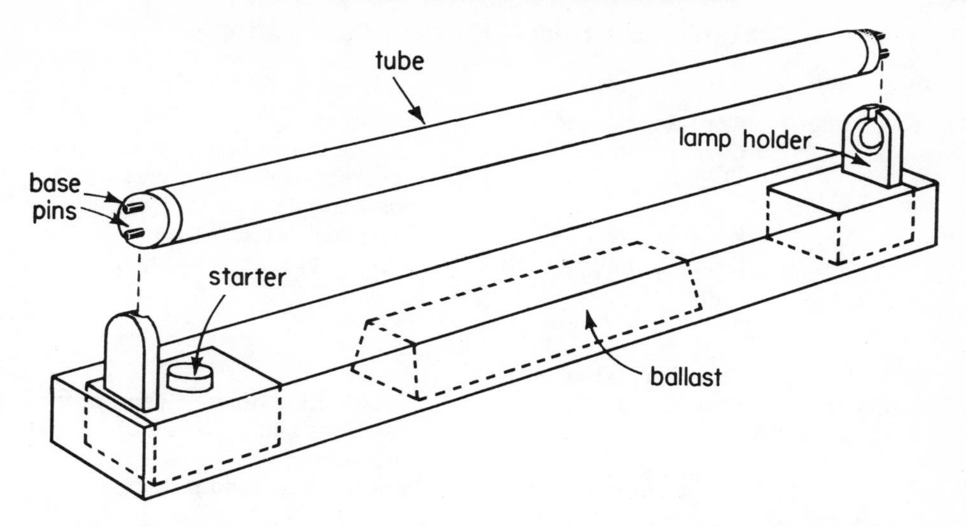

Fig. 5-10. A typical fluorescent light and how it is put together. The inside of the tube is coated with fluorescent powder for color and the light tube is filled with argon gas and mercury vapor.

plain white or silver metal housings, inside of which are the ballast and the wiring (Fig. 5-10). Two tube-holders or sockets hold the ends of the fluorescent tube. Most tubes have two prongs, or base pins, on each end, which are inserted into the holders and gently twisted about ¼-turn to make contact. The starter, if any, is like a tiny tin can, which sticks out of a hole in the housing, usually underneath the tube.

Causes of Failure

If the light does not operate, first rule out the usual causes of electrical failure—a cord (if any) not plugged in, a blown fuse or tripped breaker, defective plugs or cords. Do not suspect the bulb immediately. Fluorescent bulbs ordinarily last five to ten times longer than incandescent bulbs, and usually give several signs of dying before they actually go, such as a hum or, in the case of a defective ballast, a distinctive, unpleasant odor.

Check the diagnostic chart on page 48 for possible causes and cures of fluorescent ills. Before going to any great lengths, however, there are a couple of things you can do that are easy to find the cause of some common problems. First, twist the bulb a few times in the holders to make sure it is seated right. If the trouble persists, remove the bulb by turning it counterclockwise until the prongs line up and the bulb can be pulled out. With the bulb removed, take one socket or tube-holder in each hand and bend slightly toward the center of the fixture. Now try the tube again.

If that doesn't work, and the fixture has a separate starter, try replacing that. There are quite a few types and sizes for all fluorescent parts, so make sure you get the same type replacement starter. The same applies to all other fluorescent parts. Take the old one along if you have any doubts.

Replacement of Bulb or Starter

Replacement of the bulb or starter is a simple task. There are two other types of bulbs that are slightly different, however, from the usual two-prong long tubes. Usually not found in the home, one-prong instant-start long lamps are removed by pushing in on one end, where the spring-wound

Diagnostic Chart—Fluorescent Lamps

Symptoms	*Cause*	*Treatment*
Tube won't light	Cord unplugged	Plug in.
	Circuit fault	Replace blown fuse or reset circuit breaker. Determine cause in circuit and fix.
	Worn starter	Replace, if preheat type.
	Oxide build-up on tube prongs	Rotate tubes, sand prongs.
	Faulty tube-holder	Replace.
	Tube burned out (blackened ends)	Replace.
Flickering (constant)	Poor contact	Press in tube-holders, sand prongs of tubes and tube-holder contacts.
	Oxide buildup on tube prongs	Rotate tubes, sand prongs.
	Cold fixture	Raise room temperature, minimum 50° F.
Discolored tube ends	Defective starter	Replace, if preheat type.
	Erratic tube	Turn tube around if only one end discolored.
	Defective tube	Replace.
Unlit tube center	Defective starter	Replace, if preheat type.
	Defective ballast	Replace.
Noise, humming	Defective or wrong type ballast	Replace.
	Incorrectly installed ballast	Check wiring and correct.

tube-holder is pressed back to release that end. The other end is then pulled out. (If one end doesn't move, try the other.) Some homes may have a circular type of bulb. To remove a circular tube, carefully disconnect it from the socket, then pull it out of the retaining clips on the fixture. Connect the new tube to the sockets, then press into the retaining clips.

Replacement of Tube-holder and Ballast

Replacement of a defective tube-holder or ballast is a little more complicated, but really not difficult. Since both replacements involve the wiring, make sure that the current is off before attempting either. Take off the cover plate by loosening the screws at each end. This will expose the wiring for both tubeholder and ballast.

To replace a tube-holder, disconnect the wires leading to it by removing the wire nuts or terminal screws. Remove the tube-holder by taking out the mounting screws. You may also have to take off the end bracket so that you can slide the tube-holder off its mount. Install the new tube-holder by reversing the procedure.

The ballast is the business part of the lamp. It consists of wire wound around a steel core, which momentarily delivers a higher voltage when the fixture is turned on, sending an arc through the argon gases that produce the light. The ballast also stabilizes the lighting by limiting the total amount of power that can flow through it. As noted previously, a defective ballast often signals its end by starting to hum or, depending on the mode of failure, by giving off a bad odor.

To replace the ballast, first disconnect all its wires. This may involve loosening several wire nuts and/or terminal screws. The wires should be color-coded to prevent miswiring. In any case, you might feel safer by labeling each one with a piece of masking tape indicating to what it was attached. Hold the heavy ballast with one hand, removing mounting screws with the other. Take the ballast to an electrical supply house (about the only source for these), and get an exact replacement. Reverse the above procedure to install the new one. (A partner may help to hold the ballast while you remove or secure it.)

Replacing Ceiling or Wall Fixtures

Although the wiring within an individual ceiling fixture can be quite intricate, particularly if the fixture has multiple lights, it is usually a simple job to remove and replace any type of wall or ceiling fixture. Wiring to and from the fixture is straightforward. Any fixture you buy should come prewired, and all you have to do is make connections at the box.

Attachment Devices

No matter what type of fixture you are installing, study the old fixture to determine the mechanical attachment devices. After shutting off power to the fixture, look for the ornamental nuts in lighter weight fixtures, and the larger, center locknuts on heavier types. Remove the nuts of whatever type, and see if the fixture is loosened. If so, hold it up (you or a helper) until the wires are removed. Use a coathanger or other device to do this if necessary.

Check the mechanical attachment devices of the old fixture, as compared to the new. You may be able to use the same bracket or whatever to attach the new one, but it will probably be easier to remove the old devices in order to get at the wiring. There should be only two wires connecting the fixture to its circuit. Detach the old fixture from the wiring by removing the wire nuts or other connectors. Set the fixture aside for repair, installation elsewhere, or the garbageman.

Some fixtures will have one black and one white wire leading to it. Others, depending on the switching and wiring arrangements, may have two black wires (or a white wire painted or taped black). If there is one white and one black wire, connect the new wires accordingly, black to black and white to white. If both are black, it makes no difference. In any case, connect the wires with wire nuts, then secure the fixture as directed by the manufacturer, or by using one of the devices mentioned above.

Technique

Ceiling fixtures are ordinarily attached to a shallow round or octagonal box (Fig. 5-11). You may need a hanger strap, which bridges the box from one side to the other, a threaded stud, and/or a hickey connector as shown in Figure 5-12. (Note: This is not the same as the hickey conduit bender described in Chapter 3.) There are many different ways of making the mechanical arrangements, and the fixture supplier should provide either the needed parts or directions for making this attachment. Even if he doesn't, you should be able to figure it out by studying the fixture and the diagrams in Figure 5-12. Heavier fixtures, such as dining room chandeliers, usually

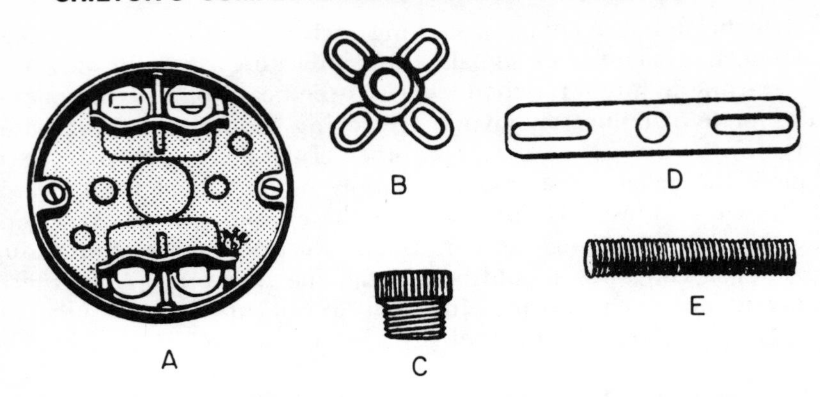

Fig. 5-11. (A) Ceiling box, (B) Fixture stud, (C) Extension nipple, (D) Strap, and (E) Nipple.

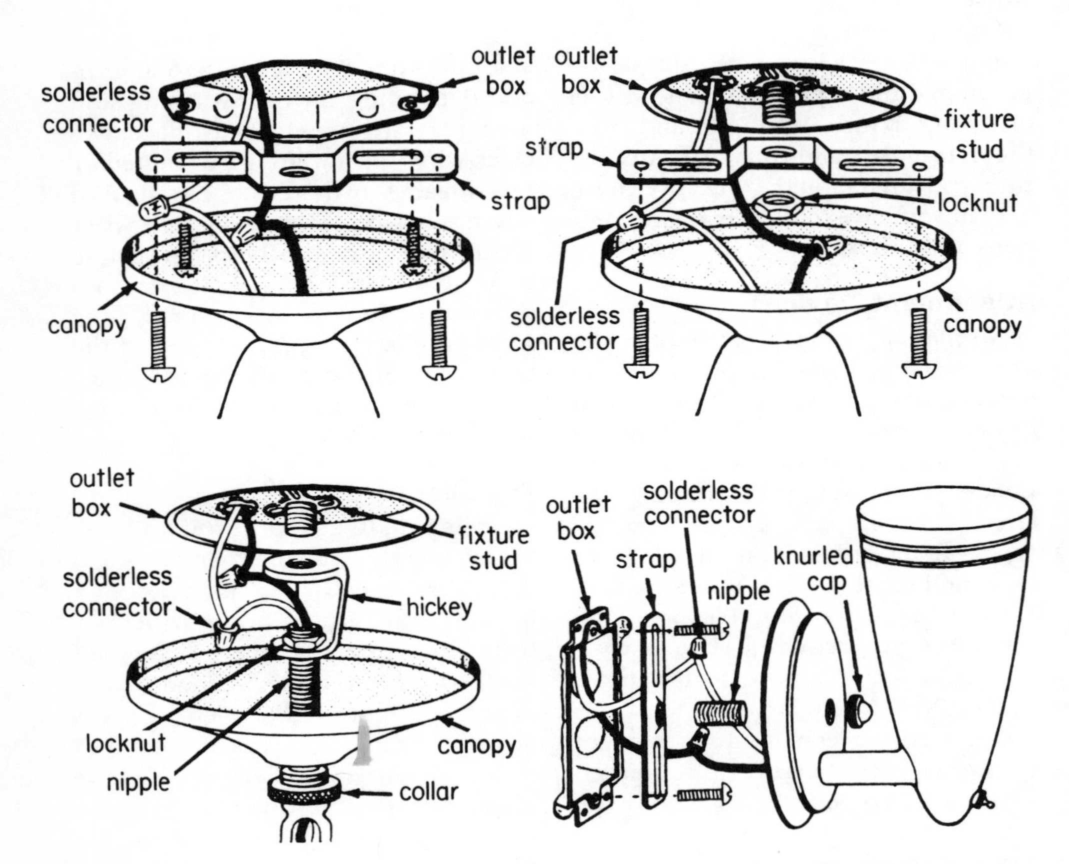

Fig. 5-12. Some of the ways of attaching ceiling and wall (bottom right) fixtures to boxes. Note the "hickey" at bottom left to attach nipple to a center stud. (Top, left) If there is no stud in the box, the strap is fastened to the threaded ears, and the fixture is fastened to the strap. (Top, right) If there is a stud, the strap may be fastened with a threaded nipple and locknut. (Bottom, left) Heavier fixtures use two nipples joined by a hickey. (Bottom, right) Wall fixture using strap and nipple.

employ the hickey-stud-nipple arrangement shown (Fig. 5-12, bottom, left), but sometimes a threaded stud is inserted into the center of a mounting bar and secured with a locknut (Figure 5-12, top, right).

You should, by the way, make sure that you have the proper connecting devices needed for the particular fixture before you start the replacement process. When you buy a fixture kit, open the box to see if all the parts are included. If not, ask the dealer to provide you with the proper parts. It's more than a little aggravating to be perched on a ladder holding up an already wired-up chandelier, only to discover that you're missing a nut or something to hold up the fixture.

You should also be prepared for what can at least seem to be a rather lengthy period of arm-tiring overhead work. It is advisable to have a helper to hold the fixture while you are making the various attachments, or to spell you. He or she can also hold a flashlight if needed. Also, be sure to have a sturdy chair or ladder to stand on.

Wall fixtures are removed and attached the same way as ceiling fixtures. Although you may not need anything to stand on, a helper is still advisable to hold the fixture while you perform the hookups.

Replacing Outlet Receptacles

How do you know when an outlet receptacle is defective? Well, you have a pretty good clue when a portable lamp goes out, and a new bulb doesn't correct the problem. It could be a defective new bulb, of course, but it is easy enough to check this by trying out the bulb in another lamp that is already working. If the bulb works there, you know it's not the problem. (Return any defective bulbs for a refund.)

The problem could be in the lamp itself. To determine this, unplug the lamp and try another outlet on the same circuit. If the lamp works, strongly suspect the outlet itself.

New receptacles are readily available in hardware stores, electrical supply houses, and many drug or variety stores. Most likely, any new one will have three holes for each outlet, the third one (at top or bottom) for the ground prong. You can use this type for replacement, even though the existing wiring is not equipped with a ground. (As discussed on p. 57, attach a ground wire to the green octagonal ground-wire screw terminals.)

Technique

First, make sure that the power is off by removing the fuse or tripping the breaker on that circuit. If you don't know which circuit serves the breaker (see Chapter 6), turn off the main fuse or breaker. Then, remove the cover plate on the old receptacle by unscrewing the small screw in the center (Fig. 5-13).

Remove the receptacle attachment screws at the top and bottom (Fig. 5-14). Pull out the old receptacle from the box and remove the wires by turning the set screws in a counterclockwise direction (Fig. 5-14). (Turn the screwdriver to the left.) Disengage the wires, leaving the loops intact.

Some receptacles may be "backwired." These are designed so that you can insert the wire through an aperture that automatically grips for a solid connection. (Fig. 5-15). To detach the wires from the push-in terminals, insert a small screwdriver blade into the special release aperture next to each terminal aperture where the wire is inserted, pulling on the wire at the same

Fig. 5-13. A receptacle cover plate is easily removed by loosening the center screw, but make sure that the power to the outlet is off.

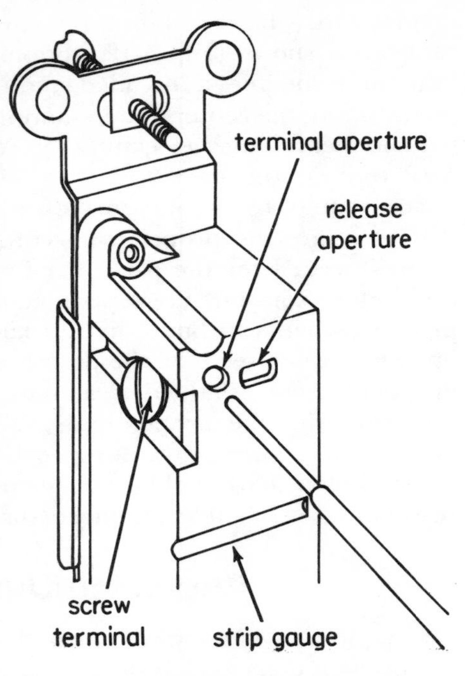

Fig. 5-15. Backwired receptacles can be loosened from the wires by pressing a screwdriver point into the release aperture.

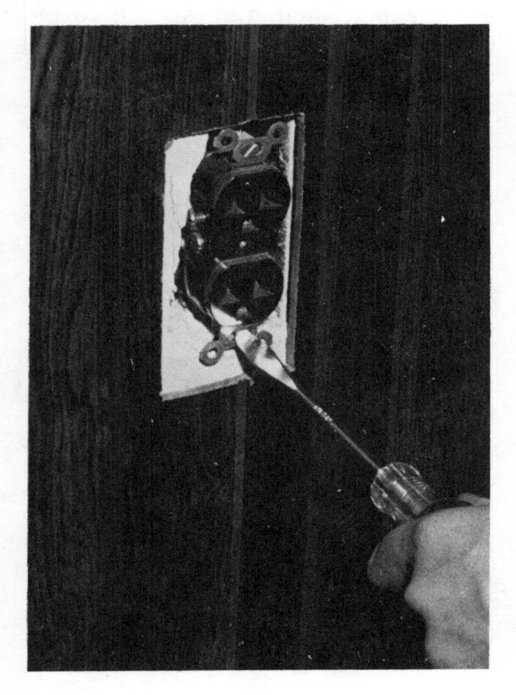

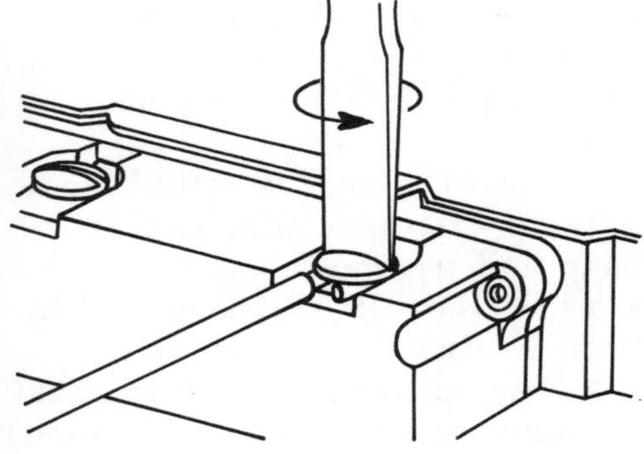

Fig. 5-14. (Left) The receptacle is detached from the box by removing the screws at top and bottom. (Above) Remove the wires from the old receptacle by loosening the screws to which the wires are attached.

time (Fig. 5-15). If the receptacle is at the end of a circuit, only one black and one white wire will be attached. Middle-of-the-run outlets will have two black and two white wires attached (one set entering and one leaving the outlets).

If your home has a relatively new electrical system, there will be another set of wires, either bare copper or wrapped in green insulation. This is the separate ground wire. Remove this, too, noting how it is attached.

When all the wires are disengaged, remove the old receptacle and throw it away. Replace all wires the same way that they were attached, white wires to nickel-colored screws, black wires to copper screws. Attach ground wires as discussed on p. 57.

Push the receptacle carefully back inside the box, making sure that no connections have come loose in the process. To avoid the possibility of a short circuit, make sure that no wire ends are sticking out and all solderless connectors (if used) are secure. Attach the mounting screws through the slots at bottom and top, straightening the receptacle if necessary by adjusting the screws in the mounting slots, moving to right or left as needed. Replace the cover plate.

Replacing a Single-Pole Switch

A single-pole switch is used when a fixture or outlet is controlled from only one point. Fixtures that are controlled from two or more places, such as stairway or hall lights, will have three-way or four-way switches. Replacement of these is a little more complicated, and pages 54–56 should be studied before attempting to do this.

Causes of Failure

A single-pole switch is defective when the fixture(s) it controls refuses to be turned on or off. If it cannot be turned off, you can be quite sure that it's the switch. If the fixture cannot be turned on, first try a new light bulb that you know is good. When that doesn't make any difference, it's a good bet that the switch is defective. The fixture may be defective, of course, but it's easier and less time-consuming to check out the switch.

You have several choices at this point. If you don't want to mess around with the testers described below, simply replace the switch. The odds are high that this will solve the problem. Switches are not overly expensive, and no harm is done if that isn't the source of difficulty.

Testers

Neon Tester

However, you can make sure that the switch is good by using a neon tester. You will have to leave the current on, but there is little likelihood of your getting a shock if you exercise a minimum of care. Turn the switch to the "on" position (up). Then, remove the switch plate by loosening the two screws at top and bottom. Place the prongs of the tester on the two dark or copper screws of the switch. (There should be no nickel-colored screws on a single-pole switch.) The screws are usually located on opposite sides of the switch, but both may be on the same side. If the switch is backwired, it's difficult to test without pulling out the switch. Just go ahead and replace it.

If the neon bulb doesn't light, when both prongs of the tester contact the terminals, flip the switch to the other position. (It may have been installed upside down.) When the neon still doesn't glow, you can be sure that the switch is defective. The tester shows the presence (if it lights) or the absence (if it doesn't light) of line voltage.

Continuity Tester

You can, of course, play it perfectly safe and use a continuity tester. To use a continuity tester, you have to deenergize the circuit and remove the switch. It seems to me that if you're going to all that trouble, you may as well simply replace the switch. Chances are that the switch is old anyway and worth replacing with a new, quieter, mercury-type switch or dimmer control (see p. 23).

Technique

In any case, when you have determined that the switch should be replaced, kill the current to that circuit. Detach the mounting screws at top and bottom, then remove the wires from the screw terminals (Fig. 5-16). Both of these should be black. If there is a white wire leading to the switch, as there sometimes is, depending on where the switch is connected to the circuit, it should be painted or taped black at the end. If that wasn't done, do it now, since all wires leading to a switch are always "hot."

There may also be two white wires

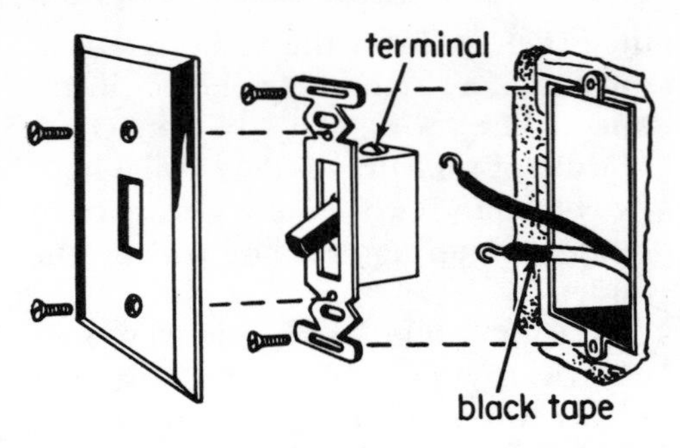

Fig. 5-16. Removal and replacement of a single-pole switch. See text for details.

inside the box, tied together with a wire nut or other connector. Don't disturb these. They are the other continuous half of the circuit. But, if for some reason, they become detached, reconnect them. If there is a ground wire (bare or green-covered), detach that also, observing how it was connected and reconnect it later the same way (see section on Grounding if there are any problems with this).

Connect the new switch to the black wires using the same techniques as for receptacles. It doesn't matter which wire is connected to which terminal in a single-pole switch. Push the switch carefully back into the box, replace the cover plate, and reenergize the circuit, following the same general guidelines used for outlets.

Replacing Three- and Four-Way Switches

Three-way and four-way switches are very handy where it is inconvenient, difficult, or dangerous to turn off a light until you've reached your destination, or when you wish to be able to turn on a light at some point before getting there. Some common uses for this type of switch are in long halls, stairways, or in a garage. You turn on the light at one point, then turn it off at the opposite end. If you couldn't do that, you'd have to either climb the stairs in the dark, or leave the light on all night.

Actually, these switches appear to be misnamed. A three-way switch controls a light or receptacle that can be turned on or off at *two* locations. A four-way switch controls the fixture from *three* points. The terms, however, refer to the number of contact points *inside* the switch, not to the number of switches used.

Principles of Wiring

The wiring of these switches seems quite complex, but the principle is simple. The switches are thrown in such a way that current can be interrupted or continued in several different ways. If you study the diagrams in Figure 5-17, you can trace the current in the four different configurations (two on, two off) and see how the switches work.

Don't feel bad if you don't get it. It's very confusing at first, although if you

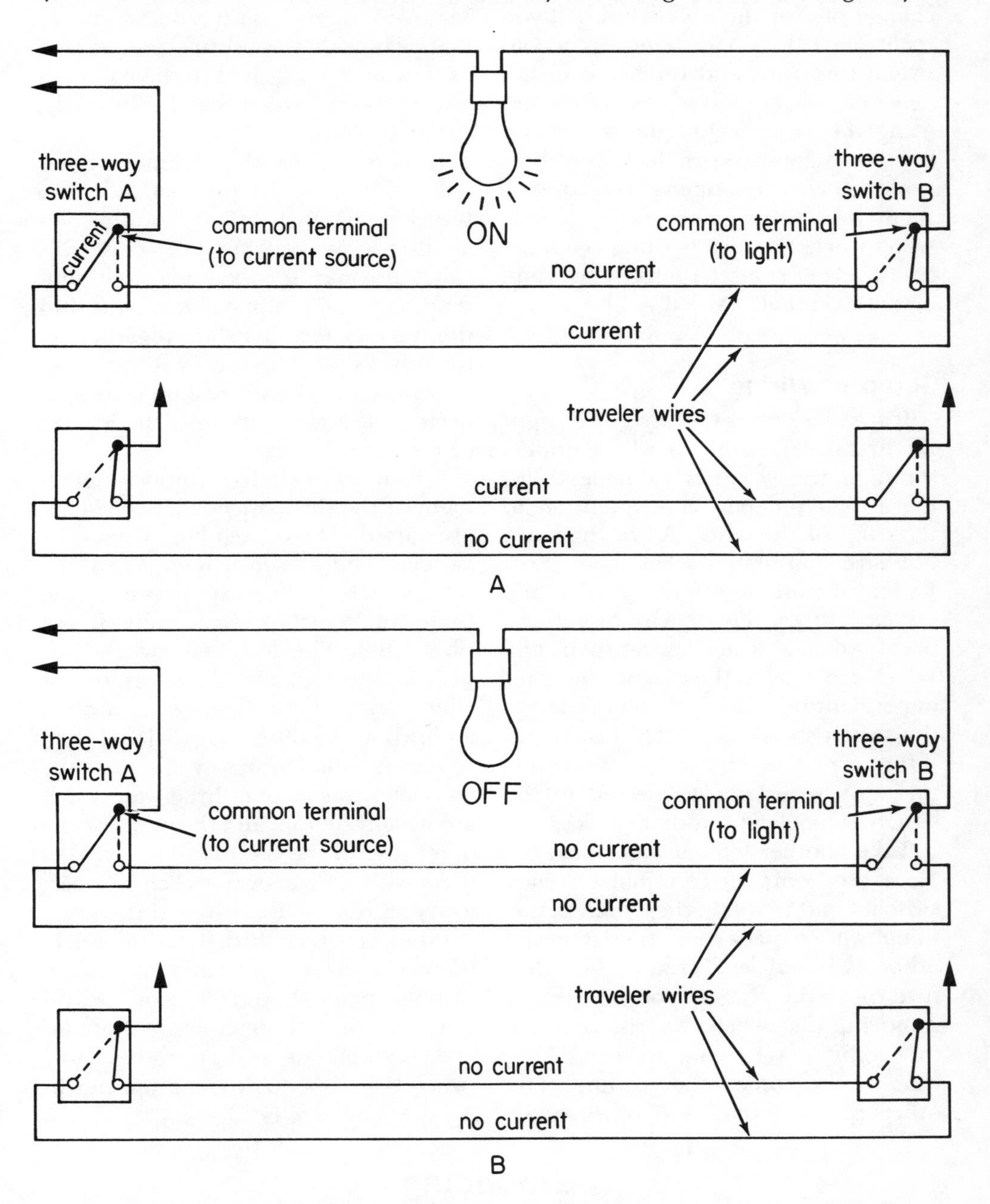

Fig. 5-17. A three-way switch can be controlled from two locations because the current can be transferred to either of two traveler wires. If the two switches are in either of the "on" positions shown in A, there is current flowing through one of the two traveler wires. If the switches follow the configurations shown in B, no current will flow through either traveler and the light is "off."

ponder it a while, you'll eventually understand and wonder why it took you so long. The point is that you don't really have to understand the theory to change one of these switches. All you really have to do is remove the old switch carefully, and replace it with a new one, wiring it as it was before and using the same technique as you did for the single-pole switch. In any case, nothing evil will happen if you connect them incorrectly. The system simply won't work. So, if something is wrong, try different combinations by trial and error. Eventually, you'll get it right.

Setup of Wiring

It's easier, however, to get it right the first time, and that's where understanding the setup is valuable. With these switches, pay little attention to the color of the wires. All of them are hot wires, no matter what their color. Red insulation is often used for the traveler wires. These wires interconnect switches when more than one switch can control the circuit power independently. White wires connected to the switch should be coded black, and if they aren't, do so yourself. Don't disturb any wires not connected to the switch. They lead to other devices.

Take another look at Figure 5-17A. Note the common terminal on each switch. On the switch, the common terminal will be black or darker than the other two, and lead either to the fixture or to the current source. Before removing the switch, it's wise to mark the common terminal with masking tape or use some other coding. The other two terminals will be brass or nickel-colored, in any case lighter than the common terminal. The wires that are attached to these are called the traveler wires. These lead from one switch to the next, and it doesn't really matter how they are connected, as long as they aren't attached to the common terminal—in which case, obviously, they won't work.

In Figure 5-17, the common terminal is shown at the top, with the two traveler terminals at the bottom. Not all three-way switches are made this way. In some, for example, the common terminal is alone at one side, and the traveler terminals are together on the other side. The theory is the same, however, and the wires should be connected as shown, with only the *position* of the wires changed.

A four-way switch is simply a refinement of the three-way. Three switches are used, two regular three-way switches and one four-way. The four-way switch is usually between the three-way switches (electrically, if not physically). The four-way switch simply contains four traveler wires, two of which lead to one three-way switch, and two to the other (Fig. 5-18). When replacing the four-way switch, the traveler wires to one three-way switch are usually on top, and the ones for the other are on the bottom.

As with a three-way switch, it's best to try to replace the wires in the same manner as before. If that doesn't work, however, resort to trial and error. The traveler pairs should be together on one side, if not together at the top and bottom. As with the three-way switches, colors don't mean too much. All the wires are hot.

Grounding

One of the dangers of electrical work is that a wire can be loosened and hot current can be transferred to an object or being that was not designed to handle such current. If that being is you, the current can flow through your body into the earth, or "ground." Thus, your body itself can act as a con-

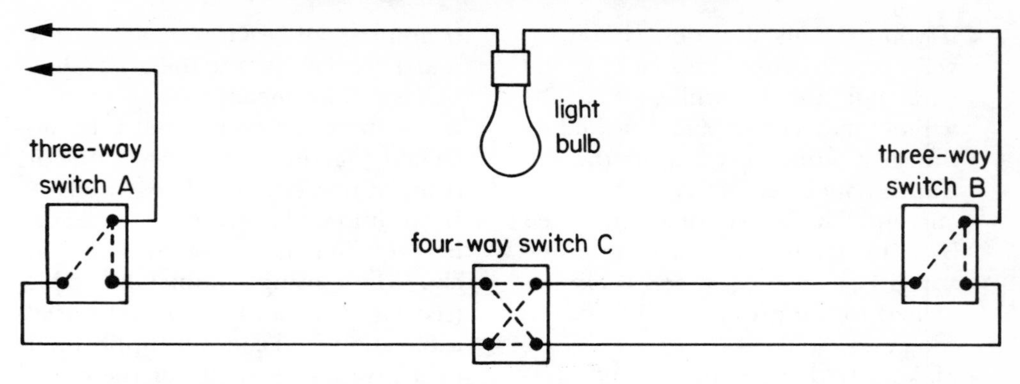

Fig. 5-18. To control a light from three or more locations, a four-way switch is used between two three-way switches. Current is turned on or off along one of the many pathways shown.

ductor, which is exactly what happens in the electric chair mentioned in Chapter 1. The severity of the shock depends, among other factors, on whether your feet are dry, damp, or wet—in ascending order.

Some of the objects that can unintentionally become conductors of hot current are the nonelectrical, metallic parts of motors, appliances, boxes, and similar items. The problem with such current leakages is that they are often too small to be detected by the fuse or breaker. However, it takes only a few thousandths of an ampere to kill a human being, which is why current leakages are so dangerous. (In a short circuit, two current-carrying wires will touch and cause a high-enough surge to blow a fuse and deenergize the circuit.)

National Electrical Code Protections

To prevent these low-resistance "fault currents," which do not blow fuses or trip breakers but can kill, the National Electrical Code has specified some elaborate and rather onerous protections. Each is designed to sidetrack these faulty currents into a safe channel. This is the reason for the ground wire from a clothes washer in an older house to the water pipe. (Newer installations have three-prong adapter plugs that are in turn plugged into a three-wire grounded circuit.) It is also the reason why the entire electrical system is grounded to a water pipe or to a pipe in the earth outside your home.

Application of the Theory of Grounding

The theory behind grounding is quite complex and is explained in more detail in textbooks. I won't discuss it here. What you should—indeed, *must*—know is how to apply it. There are just a few important applications to residential wiring:

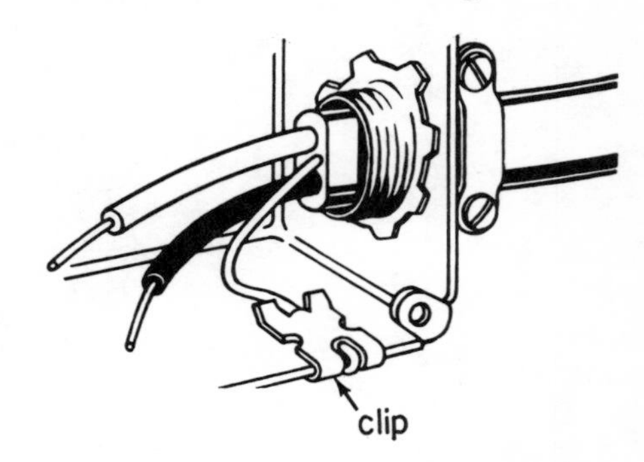

Fig. 5-19. Ground wires should be attached to each wire by means of a clip (shown) or a grounding screw. When the wires continue to another box, the ground wires are pigtailed with a jumper wire to the box.

- If you ever install a new circuit, be very sure to fuse the black or hot wire, but not the white wire. The white (neutral) wire is attached to the grounding bus bar in the entrance panel (see p. 81) so that stray currents can be eventually funneled back to earth. The bare or green wire in cable so equipped is also attached to the grounding bus bar in the panel.
- Make sure that you have an uninterrupted ground wire from the beginning of each circuit at the panel breaker back to the grounding bus bar. This means that all ground wires (bare or green) must be attached together with a wire nut or compression ring inside each box.
- In addition, the ground wire must be pigtailed to a ground screw or clip when using nonmetallic cable (see Fig. 5-19 and p. 57). Armored cable and metal conduit supply their own ground by virtue of their metallic covering.

Home Circuitry

The word circuit comes from the Latin word *circuitus* meaning to "go around." Politicians, traveling salesmen, itinerant preachers, and similar tradesmen, "make the circuit" by starting from one place, going to a few others, and then coming back home again (going around in circles, perhaps). An electrical circuit does the same thing. The current starts at the entrance panel, travels to where it is useful, then comes back again to the service.

Those who are blissfully indifferent or unaware of what happens to the electric current in their homes have little need to understand exactly what happens to the current that comes into their homes. The light goes on, the oven cooks, and the razor shaves, when the switch is on. It doesn't matter how this happens. However, if you are going to perform even the smallest electrical task around the house, it behooves you to become acquainted with electrical circuitry.

Home circuitry can vary widely, depending on the size and age of the home, the availability of other types of energy, and even the region in which you live. A small, older home in a temperate climate with a natural gas line may have only a few electrical circuits (although it could probably use more). A large, new home in the Sunbelt, where there is no natural gas, can have many more circuits. Not only would such a house have more rooms, but it would need circuits for heating, cooking, clothes-drying, and especially air-conditioning. It would also need a much bigger service (more housepower) than the former, since the number of circuits and the capacity of each would require a higher capacity entrance panel.

Determining the Number of Circuits

To determine how many circuits your home has, check the entrance panel. Under ordinary circumstances, opening up the panel door should pose no hazard, but moisture, a wiring defect, a misplaced hand, or some unforeseen circumstance could possibly be dangerous, so play it safe and stand on a dry wooden board when opening the panel door. This will prevent your body from acting as an unintentional (and deadly) ground. And, for the time being at least, don't touch anything inside the box.

Look at the fuses or circuit breakers inside. Each fuse or breaker controls a separate circuit (Fig. 6-1), with the exception of the main and certain high-wattage appliance circuits, which usually contain two linked or ganged double breakers as explained on page 61. You can determine the number of

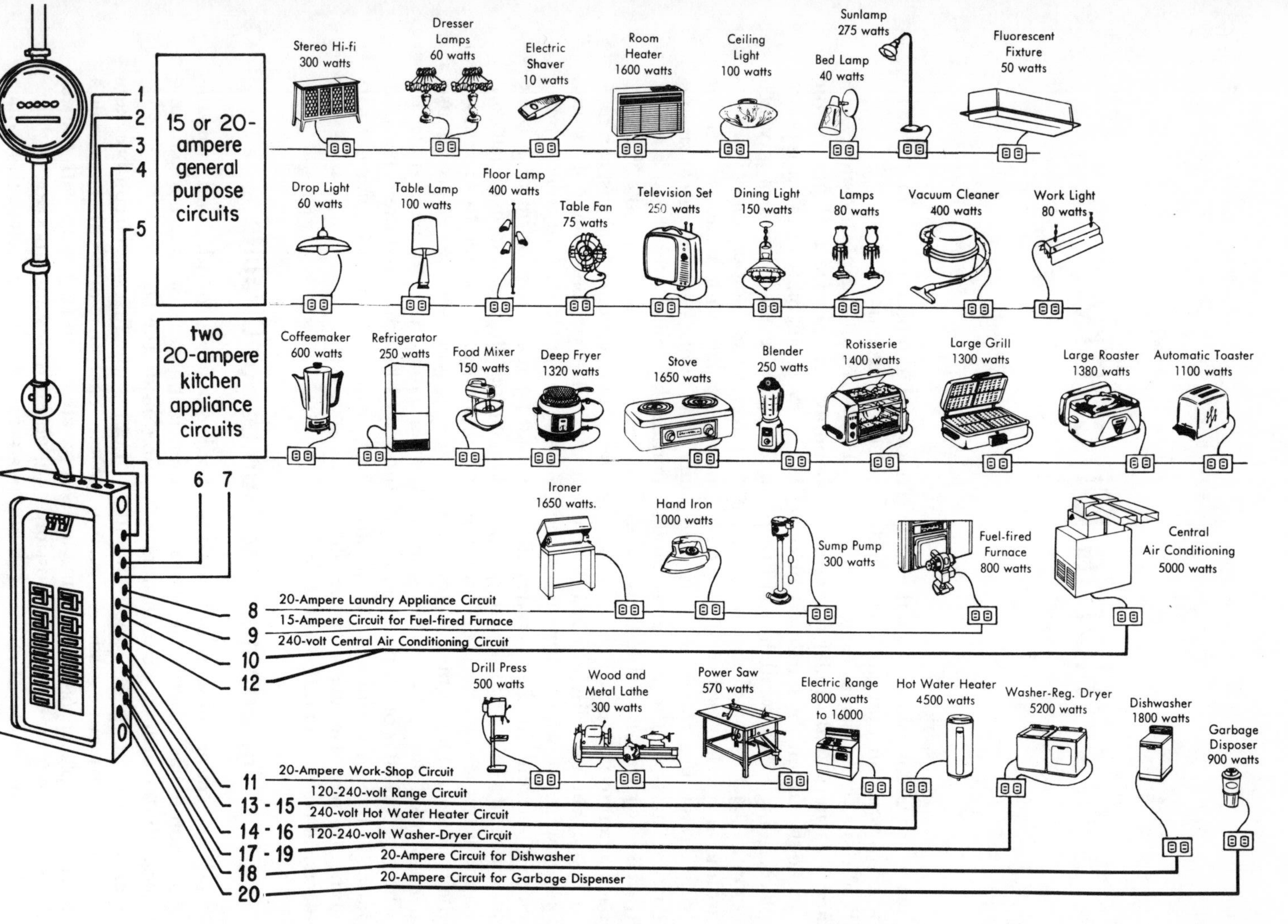

Fig. 6-1. Typical circuit setup for a home with many electrical appliances including central air-conditioning. The recommended service for this type of home is 200 amps. Note that the code now requires an individual refrigerator circuit in any new construction.

in-use circuits in your house by counting the number of fuses or breakers. Count each set of ganged breakers (excluding the main switch) as one circuit for this purpose, even though it uses two lines.

Types of Circuits

Individual Appliance Circuits

These are provided for such high-wattage appliances as ranges, central air-conditioners, and electrical dryers. These usually operate at 240 volts and are served by two lines working together. Other individual appliances use only one line, for example, oil burners, hot-water heaters, certain workshop tools, and clothes washers. Individual appliance circuits, whether they operate on 120 (one line) or 240 voltage (two), either terminate in a single wall outlet (such as an electric clothes dryer) or are wired directly into the appliance itself (as in the case of an electric range). Modern homes should also have a single outlet for the refrigerator because of current demands of frost-free features.

Regular Appliance Circuits

These are used primarily in the kitchen, but can also be found in the pantry, breakfast room, laundry, and/or dining room. These circuits are designed for the higher energy-consuming appliances, such as toasters, electric coffee pots, and similar kitchen appliances. They are wired with #12 cable and 20-amp fuses.

General-Purpose Circuits

These are found in most rooms of the house. They are used primarily for lighting and low-wattage appliances, such as clocks, stereos, TVs, and so on. Such circuits generally use #14 cable and are fused at 15 amps.

The rule of thumb is that there should be at least one 15-amp general-purpose circuit for every 375 square feet of floor space, or one 20-amp circuit for every 500 square feet. The NEC also requires at least two 20-amp general appliance circuits for the kitchen-dining area, and one for the laundry room, independent of lighting fixtures.

Tracing a Circuit

If you are contemplating an extension to an existing circuit, you must know how much wattage is being consumed on the circuit at that time. This information is also vital for determining whether or not overloading of a circuit is responsible for too-frequent fuse blowing or breaker tripping. To determine this, you must know how to trace the circuit, and know precisely which outlets and fixtures are connected to it.

Finding the Route

There are certain clues that may give you an idea of the route of the circuit. Thoughtful builders or previous homeowners may have labeled the circuits at the entrance panel. Some circuits may have a tag or paper label next to the breaker saying kitchen, workshop, or whatever. You may also be able to get an idea of the general direction of the circuit by following the route of the wires from the service. These are only *clues,* remember, enabling you to start your tracing. For example, a basement circuit may also have outlets on the first floor or an attached garage, and there may be more than just the basement circuit in the basement. Good circuitry, in fact, demands that a room be served by two or more circuits, so that the entire room

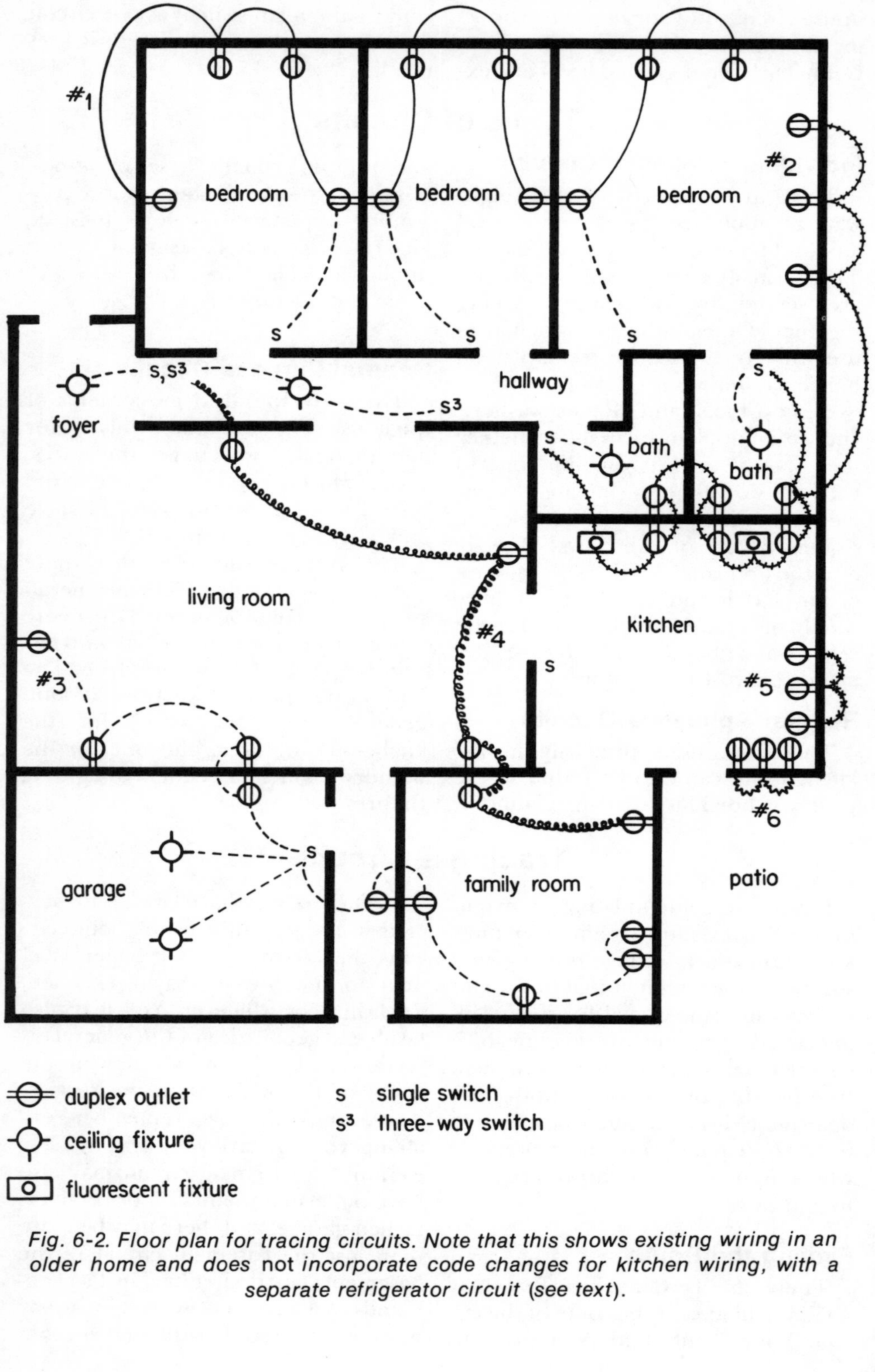
#1
#2
bedroom
bedroom
bedroom
s
s
s
s, s³
s³
hallway
foyer
s
bath
s
bath
living room
kitchen
#4
s
#3
#5
#6
garage
s
family room
patio
duplex outlet
ceiling fixture
fluorescent fixture
s single switch
s³ three-way switch

Fig. 6-2. Floor plan for tracing circuits. Note that this shows existing wiring in an older home and does not incorporate code changes for kitchen wiring, with a separate refrigerator circuit (see text).

need not be darkened by the failure of one of them.

Mapping the Route

To trace your circuits, first arm yourself with a floor plan of the entire house. A rough drawing will do for now. Mark down every receptacle, switch, fixture, and so on, in the home, using the symbols in Figure 6-2. Then turn on all the lights in the house including those plugged into receptacles. Return to the entrance panel. If it is light outside, be cautious and throw the main switch first. When it is dark, you will have to forego that step, but do be extra cautious in any case, standing on a wooden board, and using one hand to pull fuses and throw breakers. Put the other hand in your pocket, so that it won't stray against some grounding surface.

Now pull one fuse or trip one breaker, and reset the main switch. Start looking around the house for lights that are now out. Make a list of these. There will no doubt be receptacles that are not in use at the time. Using a neon tester or a small lamp (a night light is good for this), plug into all the empty outlets and make a note of all those that do not light up your testing device. Also, check all appliances, especially in older homes. Some of these may be hooked into a general-purpose circuit, not a good idea at all, but it was once accepted and quite common before the proliferation of electrical marvels.

When you have ascertained the geography of a particular circuit, trace its path on your floor plan. It doesn't matter, at this point at least, whether the routing *within* the circuit is exactly right. Switching, for example, may be a little tricky, and you may make a technical error there that can be corrected later, if necessary (see pp. 53–56). The important thing is to get all of the devices listed that are on that circuit. Incidentally, don't forget to check closets, basements, attics, and outdoor lights.

Labeling the Entrance Panel

When you have a complete tracing of the circuit, make sure to mark down which fuse or breaker controls the circuit. It is another good idea to make a drawing of the entrance panel, and to make a corresponding mark on that drawing as explained below. Also, at the same time, tag the panel itself to indicate the general course of the circuit. Gummed labels are available at electrical supply houses which can be attached to the panel. Some are preprinted with main, range, kitchen, and so on. Others are blank.

Since you will be tracing all of the circuits in the house, your map may get confusing if you link all the circuits with the same type of lines. Use col-

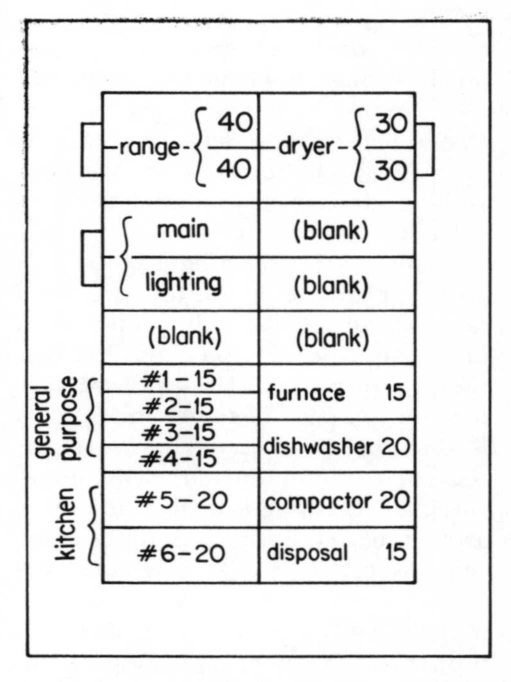

Fig. 6-3. Labeling the circuits at the entrance panel.

ored pencils, a different color for each circuit. Then, instead of continuing the line all the way back to the service, which will result in a batch of lines all running together, end the circuit line at the nearest outlet or whatever, then color-code the fuse or breaker on your panel drawing to correspond with the color of the circuit line. If you don't have colored pencils, use cross-hatched lines for one circuit, dashes for another, and so on, as shown in Figure 6-3. (Dotted lines are usually from switches to fixtures or whatever else they control.

When you complete work on the first circuit, reset the breaker or replace the fuse, after shutting off the main breaker if feasible. Trip another breaker or remove another fuse, then repeat the same procedure for each circuit until all the circuits are completely mapped out. Every outlet, switch, light, and appliance should now be located on a specific circuit.

You may wonder, in these days of energy conservation, about the wisdom of leaving all the lights on during this testing period, which can be lengthy. In spite of the wasted electricity, we feel that it's important to leave all the lights on while you trace the circuits. It is too easy to get confused if you start shutting off lights once you've assigned them to their circuits.

If, however, you've completely mapped out a certain area, such as a basement or entire second floor, and you're sure you haven't missed anything, then it's all right to douse the lights in that area. Otherwise, leave them on until you're finished. You may need the light, for one thing, and light bulbs don't really consume that much power.

Computing the Wattages

When you have your circuit map complete, the next step is to figure how much wattage is being consumed on each circuit. In this way, you will be able to determine which circuits are at or near capacity (or even overloaded), and which circuits, if any, are underutilized and can be added to.

Wattage is listed on light bulbs and most appliances. It's on top of incandescent bulbs, and usually on the label of an appliance. If you can't find the wattage for a particular appliance, use the table on page 66 to get an approximate figure. Add up the figures to get the total consumption of the circuit, assuming everything is turned on at the same time. (It may seem unfair, or unrealistic, to assume this, but see below.)

Getting a true total wattage can be a little tricky. Assume, for example, that one circuit contains four 75-watt table lamps, a 150-watt ceiling fixture, a 300-watt television set, and a pole lamp with four 100-watt bulbs. If all these are turned on at once, that's 1,100 watts, well within reason on a typical 15-amp, 120 volt circuit (multiply 15-amps by 120 volts and you get 1,800-watt capacity). But what happens if you vacuum the room at the same time? Add 400 more watts, bringing you to 1,500 watts, still okay. But also add on an air-conditioner in summer, and you may be getting up, to or beyond, the capacity of the line. (It depends on the size of the air-conditioner.)

Of course, realistically, it is highly unlikely that you will have the vacuum, television set, and the air-conditioner all working at the same time (or that all the lights will be on at once). How do you figure the usual capacity? The practical way is to add the air-conditioner, which may be in use for a considerable period in the hot months, but ignore the vacuum, or assign it to another circuit, since odds are heavy that the vacuum will be used only intermittently, and that most other devices will not be in use at that time.

You may add that it's silly in any case to use all the other wattages, because you know that you will never be using them all at once. But you don't *know*. And you must play it safe. I lived in an older home some years ago, and never had any problems with most of the circuits. We threw a large party one evening, with all the lights blazing. The blender, of course, was working feverishly on liquid refreshment. Someone decided, for a now-forgotten reason, to make toast in the dining room. The party was thrown into instant darkness.

Strange things do happen. Who knows why a teenager would want the stereo and the TV going at once? Believe me, they do. So add up your circuits assuming that all the lights and appliances are in use simultaneously. But use your head. You can, for example, assume that a portable electric heater won't be turned on at the same time as the air-conditioner.

It is easy to determine the capacity of a circuit. Each fuse or breaker is clearly marked as to the number of amperes on that circuit. Most homes have 120 volts delivered by the power company. A 15-amp circuit, indicating #14 wiring, is capable of delivering 1,800 watts (multiply 15 amps times 120 volts). A 20-amp fuse or breaker, which means #12 wiring, has a capacity of 2,400 watts (20 times 120).

What Does It All Mean?

Now that you have all the circuits mapped out, and have figured out the maximum wattage on each, what good is all this information? There are several uses to which this data can be put. For one thing, you may discover that one or more circuits are in danger of overload. To prevent this, it may be possible to shift some of the load to other, underutilized circuits. You may, for example, have a TV and a stereo on the same line. Perhaps a mere shift of a plug can produce more equal circuitry. You may wonder which outlet, or window, to use for a room air-conditioner. Put it on an underworked circuit, not one that is close to capacity. (You should also try to find a window away from the sun, but that's another story.)

The same applies to kitchens. If you have a toaster (1,100 watts), an electric coffee pot (600 watts) and a deep fryer (1,320 watts) on the same circuit (3,020 watts total), which is probably rated at 2,400 watts, you are obviously in trouble if they are all used at once. Switch them around.

Another benefit of having a circuit map is that if repair work is needed, such as replacing a defective switch, you won't have to spend time in the future figuring out which fuse or breaker controls the switch. If you keep your map handy—hanging it near the entrance panel is a good idea—you will know instantly. But always test the device as explained in Chapter 5, before plunging in to work.

For those who are planning new electrical work, circuit mapping is not only valuable, but essential. A quick look at the map and its circuit consumption will easily determine whether or not a particular circuit can accommodate additional wattage, and where prospective taps from the circuit can be accomplished (see Chapter 7).

Circuit mapping is also essential for working with old construction. Instead of having to start from scratch tracing the circuitry, this aspect of the job is already done. More on this in Chapter 8.

A circuit may go dead for a number of reasons. The first step is to determine the path of the dead circuit; it should be relatively simple to check out the inoperative lights and outlets. To determine what went wrong, try to find out where the circuit begins and start your check at that point. In most cases, a loose connection or defective device will be the culprit. The cable itself is rarely at fault unless it is old and in poor condition.

Before looking inside a box, use your neon tester to make sure there is no current to the box. Remove the outlet, switch, or lamp and examine all connections. Tighten terminals while the device is out. When you think you have found the problem, fix it and reset the breaker or replace the fuse. If the circuit goes dead again, keep checking until you've completed the circuit. If the problem persists, it's time to bring in a professional.

Power Consumed by Appliances

(Average)

	Watts		*Watts*
Air-conditioner, room type	800–1,500	Mixer, food	150
Blanket, electric	175	Motor, per hp	1,000
Broiler, rotisserie	1,400	Oven, built-in	4,000
Clock, electric	2	Radio	75
Coffee maker	600	Razor	10
Dishwasher	1,800	Range (all burners and oven on)	8,000–16,000
Dryer, clothes	4,500	Range, separate	5,000
Fan, portable	175	Refrigerator	250
Freezer	400	Roaster	1,380
Fryer, deep-fat	1,320	Sewing machine	75
Frying pan	1,000	Stereo	300
Garbage disposer	900	Sun lamp (ultraviolet)	275
Heater, portable	1,200	Television	250
Heater, wall-type permanent	1,600	Toaster	1,100
Heat lamp (infrared)	250	Vacuum cleaner	400
Heating pad	75	Waffle iron	800
Hot plate (per burner)	825	Washer, automatic	700
Iron, hand	1,000	Washer, electric, manual	400
Iron, motorized	1,650	Water heater, standard 80 gal.	4,500

7

Working with New Construction

Wiring in new construction is considerably easier than messing around with existing walls and ceilings. It's perfectly safe, too, as long as you stay away from the power source. There are no existing wires to locate, no elusive circuits to trace, no hidden framing or ductwork to hinder your progress, no need of "fishing" wire. When building an addition, finishing a basement, or whatever, the rough wiring is completed before the finishing materials are attached to the framing (Fig. 7-1). The studs and other framing lumber are in the open, easy to drill through, nail to, and pull cable through. Instead of having to fish wire through ceilings, floors, and walls, the cable can be run through or around the framing members. If there are any obstacles, such as heating ductwork, you know exactly where they are and can readily figure out how to get around them.

Planning

If your new addition or finished room is small, and there will not be a heavy wattage demand, you most likely can find an existing circuit that will handle the added load. But, if you have the capacity in the entrance panel for another circuit or two, you may as well plan an entire new circuit for the new construction.

Always use a new circuit for a new kitchen or workshop. At the other extreme is a sitting room, where just a few receptacles for lighting are planned. Most existing circuits can handle another outlet or two, perhaps more (see Chapter 6).

Extending Circuits

When you have your construction plans drawn up for the new work, determine the wattage needed, using the methods described in Chapter 6 and find the circuits closest to the new work. If you have not already estimated the load on the nearby circuits, do so now. See how much extra capacity is available, and if the new work can safely be added to the present load.

Use your imagination when planning, trying to allow for every contingency. Perhaps you are planning a small family room, for example. Will there be a deck, patio, or other exterior amenity? Then, how about outdoor lighting—if not immediately, perhaps in the future? Will you barbecue there? Certainly, a few extra outlets will be handy—for an electric spit, fire starter, lots of things. Remember that any heat-producing appliance, such as a fire starter uses lots of power. Better plan on a whole new circuit, if such high wattage is in the future. And make it a 20-amp circuit,

Fig. 7-1. Another good place to try your hand at new wiring is in a basement, such as the one shown here, before and after renovation. (Courtesy of Masonite)

with #12 wiring. (See Chapter 10 for more on outdoor wiring.)

Repairing Circuits

Don't just think of the new addition. As long as you're knocking out walls and things, you should plan on improving the circuitry in the vicinity of the new work. If a switch is now in an inconvenient place, this is the time to move it. Or perhaps you prefer to add an extra switch, so that the light can be turned on at two—or three—places. Stairways and long hallways are excellent candidates for three- and four-way switches.

If possible, it is wise to have more than one circuit serve a room. This is difficult to achieve in existing rooms, but you can and should try to plan that way for new work. Perhaps you can combine a new circuit with an extension of an existing circuit. Run the new cable along one side of the room, then tap into an existing circuit for another wall. That way, if a fuse blows on one circuit, there will still be some light in the room so that you can find your way around until the problem is corrected.

Placing Outlets

If you have a reasonably exact idea as to where the furniture will be located in your new room, and future relocation is unlikely, spacing of the outlets can be varied to fit the situation. In general, however, space outlets according to the following guidelines (equally apart as far as possible). Even if you do vary from this, don't leave more than 12 feet between each outlet. Twelve-foot spacing allows lamps with their usual 6-foot cords to be placed where needed and still reach an outlet.

- General-purpose outlets should be placed within 6 feet of lamps, television sets, and other small appliances (Fig. 7-2A).
- There should be one outlet at least every 12 feet on every wall.
- Outlets for general use should be about 12 inches off the floor.
- Switches should be 4 feet from the floor and within 6 inches of doors and archways (Fig. 7-2).
- Kitchen receptacles are placed about 12 inches above countertops, or 4 feet off the floor (Fig. 7-2B).
- There should be a kitchen outlet at least every 4 feet in the work area.
- Outdoor wiring must be channeled through a groundfault interrupter (GFI). (See Chapter 10 for more on outdoor work.)
- Install fluorescent fixtures wherever possible for softer, less energy-consuming light.

When building a family room, remember that you will probably have a television set, perhaps with a rotary antenna, requiring two outlets at one spot. There may also be a stereo, room air-conditioner, or other power-users in the same area, so space outlets close together, or consider multiple outlets in one spot, requiring two or more ganged boxes. Be generous with your outlets when putting in new work. It is simpler, easier, and less expensive to do it now than at a later date.

Use similar foresight in all your planning. Don't skimp on outlets just because you can't see a need for them in the immediate future. Think of the builder who planned your house originally. There was no way for him to predict a home with multiple TVs and stereos, videotape recorders, individual hair-dryers and hair setters, microwave ovens, and the other electronic marvels found in our homes today. Who knows what the future holds?

Placing Switches

Wall switches are another often neglected planning area. Just because you have no wall or ceiling lights

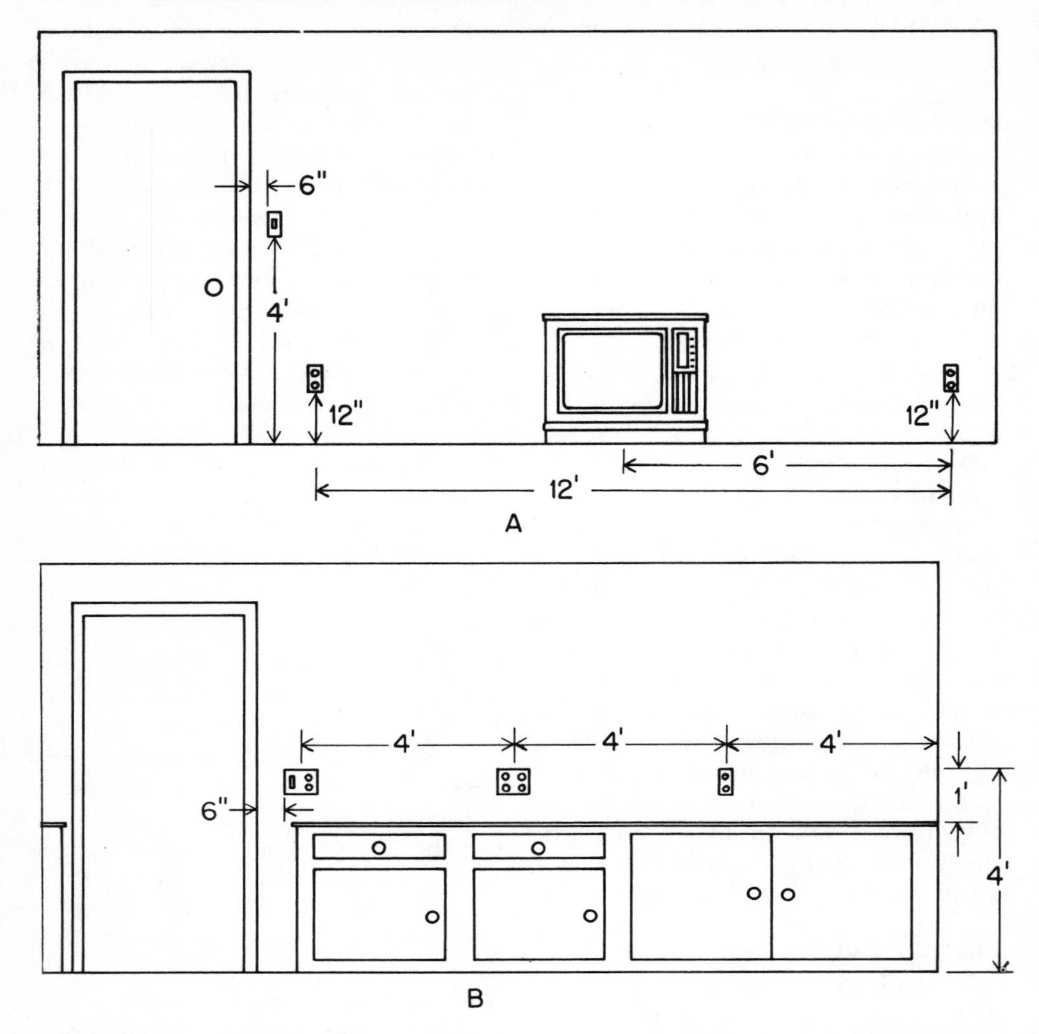

Fig. 7-2. (Above) Maximum distances for placement of general outlets and switches. (Below) Maximum distances for placement of kitchen outlets (and switches).

doesn't mean that you don't need switches. Even when all lamps are the plug-in type, switches near room entrances are always a good idea for any room. These switches can control one or more outlets in the room and prevent a lot of fumbling in the dark. In a bedroom, for example, connect a doorway switch to an outlet near a dresser or vanity. That way, you can flick on the light as soon as you enter the room rather than stumbling about trying to find the lamp. In many parts of the United States, codes require certain outlets to be connected to a wall switch by a door. Use the same reasoning for outdoor lights. An indoor switch can turn on patio or driveway lights and prevent falling over the garbage or down the steps.

Placing Lighting

Before you finalize your planning, read Chapter 11 on lighting principles. You may decide to use some of those ideas to light your own new work. Intelligent lighting can do wonders for a room that may otherwise be dull and gloomy. There is a lot more to working with light than some unimaginative builders realize. Ask anyone who works with photography, the movies, or theater.

You can do all the planning yourself, if you understand circuitry (Chapter 6). To be on the safe side, though, check with an electrician or contractor before going ahead with your plan. Be sure to consult local codes, not only for the electrical work, but for other phases of the job. If changing the home's exterior, you'll probably need a building permit.

Adding Onto an Existing Circuit

If you have determined that you can safely add onto an existing circuit for your new work, the next step is deciding where to tap into the circuit (Fig. 7-3).

The best place is at the end of the line, but this may be very difficult, if not impossible. You can find the end of a circuit by testing, as described in Chapter 6, or by simply looking for a nearby outlet with only one pair of wires, a white and a black. If there are two pairs of wires, one leading out of and the other into the outlet, that receptacle is somewhere within the circuit.

Sometimes the route from the end of an existing circuit to the new work is too circuitous to warrant tapping in there. Perhaps another circuit can be tapped into, or you may find a junction box that can be utilized. It may be possible to add a new junction box in the old line, but that may not only be a code violation, it is also probable that there won't be enough slack in the line to wire the new junction box properly.

It is possible, but generally undesirable, to begin the new wiring at a switch. This usually means that the switch will not only control whatever it

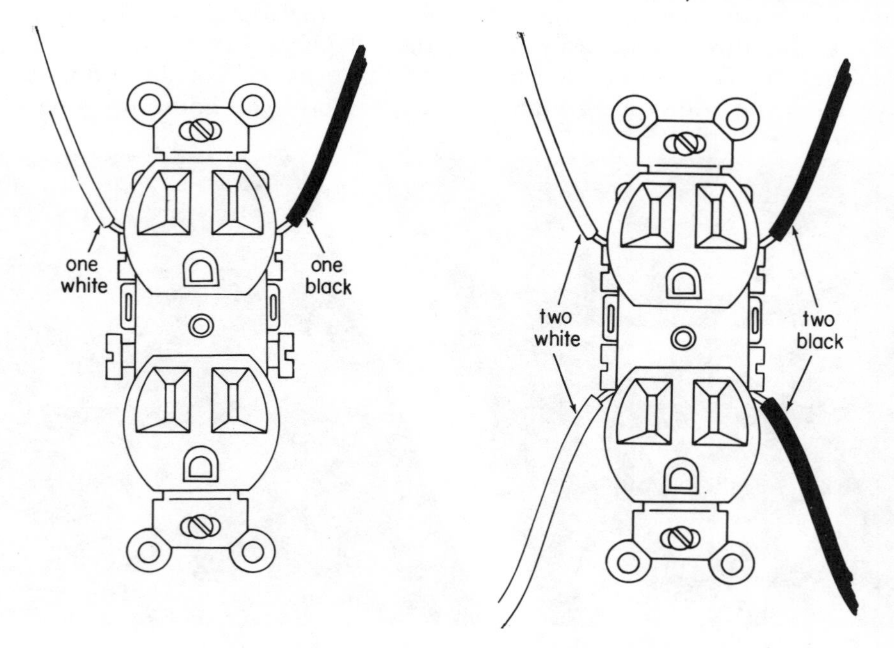

Fig. 7-3. How to attach wires to outlet receptacles (Left) end-of-circuit, (Right) within circuit.

presently operates (usually a light fixture), but also all the current in the new room(s). Sometimes this is preferable to other solutions, but it is not a happy resolution of the problem. It may be, depending on how the switch is wired, that you can tap off the switch without such problems, or that you can rewire the existing circuit to eliminate the switch control over the new work. Check all the possibilities before making your choice, and consult an electrician if necessary.

It is unwise and against codes to use an existing middle-of-the-line receptacle as a tap for new work. This necessitates jamming three cables into the space meant for two. You will have to attach two wires to one mounting screw, taking a chance on the wires slipping off. (Also see page 85 for maximum connectors in one box.) You could eliminate one outlet, and put a junction box in its place. Or, better yet, replace the present box with a larger one. It depends on how vital the existing outlet is. It may be simpler to start an entire new circuit at the entrance panel. Again, professional advice may be in order if tapping into an existing circuit presents too great a problem.

Putting in a New Circuit

If, for one reason or another, you have determined that an entire new circuit is the best choice, the problem of tapping in is eliminated, but you may wind up with another headache—getting the cable to the entrance panel. This is generally not a problem for first-floor additions or basement renovations. In most such cases, the cable can be strung through an easily accessible basement. The new cable can go down from the addition through the first floor into the basement, and from there right to the panel, following the route of the other wiring. The wiring is generally quite visible there, running along the sill plates of the house just above the foundation.

Wiring a new basement partition follows a similar path, through the drilled studs to the foundation wall, up onto the sill plate, and from there to the entrance panel. For an addition to a Cape Cod-style house or one-story

Fig. 7-4. Run the cable from the first floor down to the basement, over to the sill plate, and from there to the entrance panel.

home, the cable is strung through the joists above the ceiling to a wall near the entrance panel. You will have to "fish" (pull) the cable briefly down through the wall into the basement (or wherever the panel is located). (See Chapter 8 for details on fishing wire through old work.)

A more serious problem arises when the new work is located far from the entrance panel, and there is no clear path between them. For instance, this may happen when you add a dormer to a Cape-Cod-style house, when the rest of the second floor is already finished. It can also be a problem for houses built on a concrete slab, where there is no access below the main floor. Such problems are covered in Chapter 8. Be sure to have your cable routing worked out before starting any new construction.

Roughing In

Both professional electricians and plumbers do their work in two stages. After the framing of a new house (or addition) is completed, the basic electrical work is done. It's called "roughing in" the new work, and consists of putting up the required boxes and stringing the cable. Cabling to switches, receptacles, and so on is brought to the points where the new switches, receptacles, and so on will be located and left hanging (with enough length to ensure connection) to await finishing. After the wallboard and other finishing materials are attached to the framing, the switches, receptacles, lights and other electrical fixtures are installed.

You should work the same way. As soon as you have the framing of your room in place, the first step in your wiring job is to install the boxes. After they are put in place, the cable is then strung between the boxes. (Actually, you can put up one box at a time, adding the cable as you go. It's the same process, either way.) Complete all new wiring before hooking up to the current source.

Install the boxes as explained on pages 32–35, using the right-sized boxes to avoid Code violation (p. 85). Standard 2x3x2½-inch switch, Gem or device boxes could present a problem if you install a series of wall receptacles, in which outlets in the middle of the run contain cable both entering and leaving. To get around this, either use larger boxes, square boxes, or forego the simpler interior cable clamps and substitute exterior connectors.

Boxes should be installed so that the outside edges are flush with the finishing material (Fig. 7-5). Presumably, you know what materials you will be using, but make sure that you know the thickness of any paneling or wallboard you'll be putting up. If you haven't checked into it, you may not know, for example, that most plywood

Fig. 7-5. Positioning box for cable.

paneling manufacturers recommend that gypsum wallboard be used as a backer under paneling. In that event, add the thicknesses of both materials, and install the boxes so that they extend the same distance beyond the edge of the stud. (For ⅜ inch drywall plus ¼ inch paneling, for example, the box should stick out ⅝ inch from the front of the framing.) The Code does allow boxes to be as deep as ½ inch behind the surface of noncombustible materials, such as gypsum wallboard, brick, or concrete block. This could present a problem, however, lining up the receptacle and cover plate, so have the outside edges flush with the finishing material in all cases.

Stringing NM Cable

Unless you have a compelling reason for doing otherwise, use nonmetallic Type NM cable. It is easier and less expensive to use, and perfectly safe for any permanently dry, interior use. There are several ways of installing Type NM cable in new work, but the best way for most jobs is to drill ⅝-inch holes through the centers of the joists and studs of the framing when it runs at right angles, and to clamp it along the studs or joists for parallel runs (Fig. 7-6). Use a brace and bit or an electric drill.

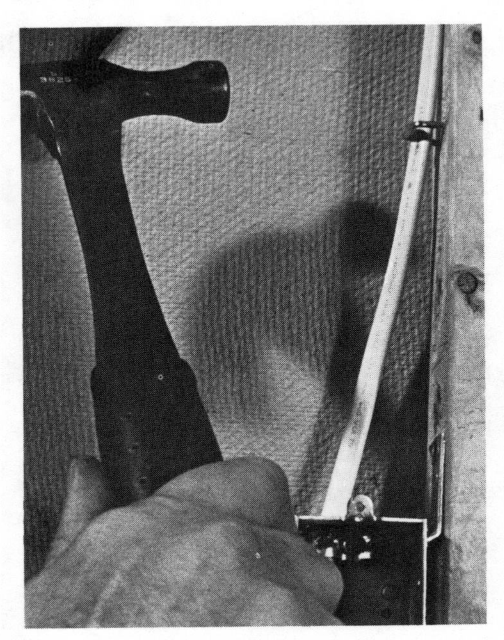

Fig. 7-6. The best way to run Type NM cable across studs and joists is to drill ⅝-inch holes through the centers of the framing. When the cable runs parallel to the framing, use cable staples to secure it. Don't hammer the staples in so tightly that they damage the cable.

In new work, getting a source of electricity can be tricky, particularly if you're building a brand new home. Your utility company may be able to run in a temporary power source. If not, use the brace and bit, preferably with a ratchet for drilling in tight corners. You'll be surprised at how effective it is.

For additions and other work where there is a power source nearby, an extension cord will do the trick. Be sure, however, to get a heavy-duty cord that is rated for use outdoors or with power tools. You will also find the cord handy for other construction work, such as cutting lumber with a power saw.

Supports and Protection

When you run the cable through drilled holes in the framing, no straps or staples are needed for support. Cable running lengthwise along joists or studs must be secured every 4 feet or less and within 12 inches of any metal box. Use small straps or special staples designed for use with this type of cable (Fig. 7-6). Do not use staples that are designed for use with Type AC cable (Fig. 3-13, right).

Staples

The advice is often given that you shouldn't use staples at all with Type

NM cable, but chances are that you'll have difficulty finding the special straps that are often recommended. Almost every supplier stocks the flattened metal staples shown in Figure 3-13, left, and it is a rare professional electrician who doesn't use them. Be very careful, however, when you drive in these staples. Don't drive them in so hard that they crush or otherwise damage the insulation. Tap them in only until the cable is secured. There is no need to grapple them to the wood as if it were going to run away.

Metal Plates

The NEC also requires that metal plates be used in the front of any framing where the holes are located less than 2 inches from the front edge of the finishing material. If you drill a hole in the center of a 2x4-inch stud (which is actually 1½x3½ inches), its center will be only 1¾ inches from the front of the stud. When you add the finishing material, though, this places the hole just about 2 inches away, maybe a little more.

The idea of all this is to protect the wiring in case someone drives a nail through later on. This is one instance when I recommend being a little more cautious than the Code (which is pretty cautious). I would use these plates whenever drilling through a stud. They should be at least 1/16x¾ inches and are available at electrical supply houses (see Fig. 7-7). Nail them to the front of the stud wherever the cable passes through. (There is no need for plates in a joist, which is always a 2x6 or larger, placing the hole at least 2¾ inches from the bottom of the joist.)

Fig. 7-7. Whenever the cable is less than 2 inches from the front edge of the finishing material, it should be protected by hammering special plates to the front of the stud.

Running Boards

Another method of stringing cable at right angles to joists in exposed attic floors or basement ceilings is to use running boards (Fig. 7-8). Pieces of 1x3 inch furring lumber are nailed to the joists, and the cable is attached to that in the same way as it is for parallel runs. This may be a good idea for attics, where insulation between the joists makes drilling difficult, but not so great for a basement, where you may decide that you want to finish the ceiling. The cable may then have to be rerouted.

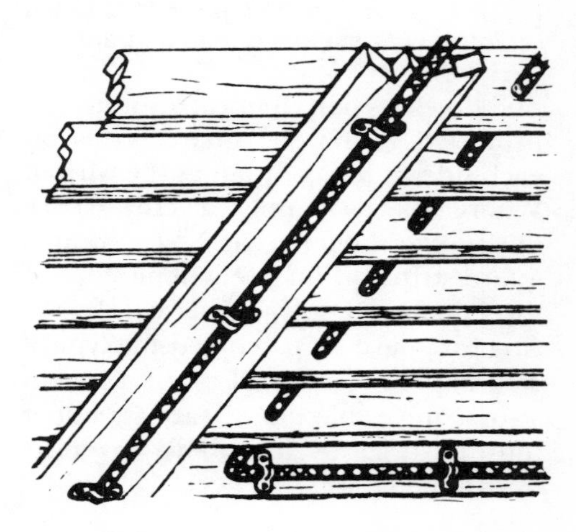

Fig. 7-8. Running board (left). Cable is run along 1x3 furring strips. At right, cable was run through drilled holes.

Guard Strips

Actually, in an attic, you can attach the cable directly to the top of the joists without a running board (Fig. 7-9). If the attic is accessible by means of stairs or a permanent ladder (including the pull-down type), the cable must be protected with wood guard strips on each side, at least as high as the wiring. Where there is only a crawl-space opening to the attic, the NEC requires guard strips within 6 feet of all sides of the opening (Fig. 7-10). Use 1x2 furring for guard strips when using typical #12 or #14 residential cable. You can also string cable along rafters, but it must be at least 7 feet above any joists over a living area.

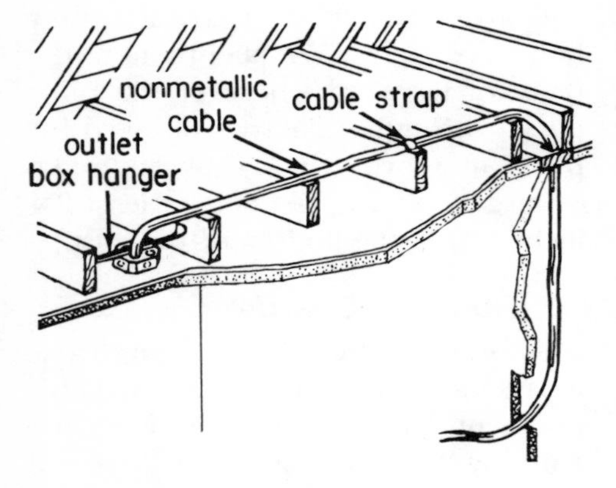

Fig. 7-9. Cable attached to the tops of the joists with staples or cable straps.

Summary

If all this sounds rather complicated and difficult to remember, I agree. Furthermore, if you think you may want to finish the attic, or part of it, someday, you will probably have to do the work all over again. Although it means a little more labor, it is simpler and often easier in the long run to follow our original suggestion—in all cases drill through the framing when working at right angles, and staple to the inside of the studs and joists when running parallel to them.

One further caution. Be sure to run the cable *over* the headers (top board) of door and window openings, not through the openings themselves. (Fig. 7-11). (Don't laugh, it happens.) Also, when making turns with cable, don't form too tight a bend. The Code says that the radius of a bend, if it were to form a complete circle should be at least 10 times the diameter of the cable. Since this is a little difficult to figure, just remember not to bend the cable so tightly that you crimp the outer covering. Keep bends as wide as you can to avoid this, preferably using a 12-inch radius or more.

When you get to each box, make connections as discussed on pages 35–38. If you prefer to strip the cable later, allow 8 inches of cable protruding into the box. As mentioned

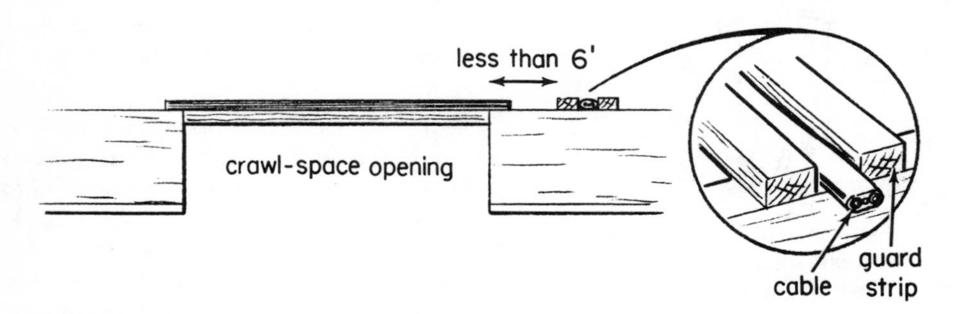

Fig. 7-10. A crawl-space opening, with guard strips placed only on cable within 6 feet of the opening.

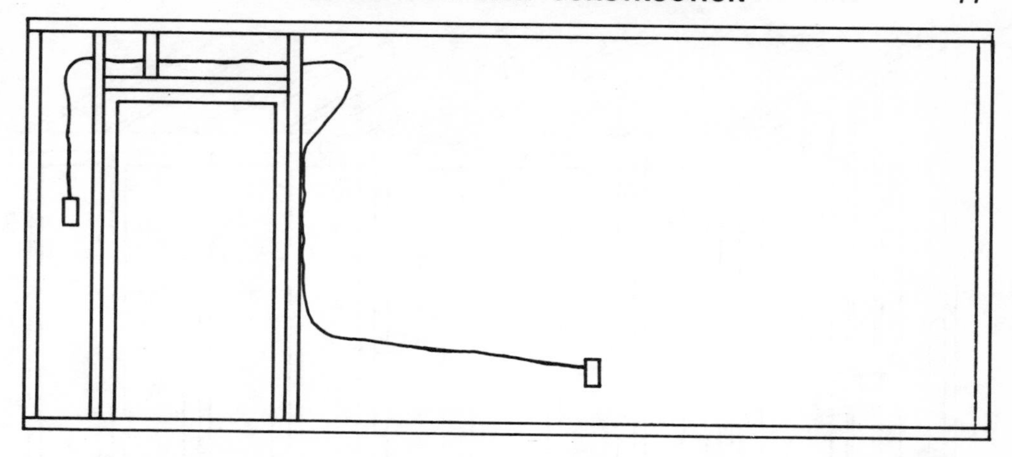

Fig. 7-11. Be careful not to run the cable through any openings for doors and windows. It should be strung around the openings as shown, with the radius of the bend at least 10 times the diameter of the cable.

previously, most of us find it easier to strip the cable before attaching to the boxes, in which case 6 or 7 inches should be sufficient.

Using Other Types of Cable

Sometimes, local codes require the use of armored cable (Type AC) or thin-wall conduit (electrometallic tubing or EMT). The National Electrical Code has no such requirement. This is an unnecessary and burdensome restriction, but you violate it at your peril, even though few people wind up in jail for this. So here's how to do it if you must.

The general method for stringing armored cable is the same as for nonmetallic cable. Armored cable is supported by either two-hole straps or special staples, as shown in Figure 3-13. Staples must have a rustproof finish to qualify for use under the Code. The supports must be used every 4½ feet and within 12 inches of each box.

Thin-wall Conduit

Thin-wall conduit also follows many of the same general principals as Type NM cable. It is, however, much more difficult to install. Conduit can be run through drilled holes. But this is usually a difficult procedure except on long runs. (It comes in 10-foot lengths.) The most generally used method is to cut notches in the studs deep enough to hold the conduit, then cover with steel plates. (Fig 7-12). Several types of straps are used to support conduit for parallel runs. They should be used every 6 to 8 feet.

Conduit Boxes

Conduit boxes must be steel or aluminum. They are always put up first. The conduit is then installed and attached to the boxes by means of special connectors (Fig. 7-13). For runs over 10 feet, couplings such as those shown in Figure 7-13 are used. Since thin-wall conduit is not threaded, as rigid conduit is, the end of the conduit is inserted into the threadless end of the special fitting, then the threaded end is turned into the box knockout and tightened with a locknut.

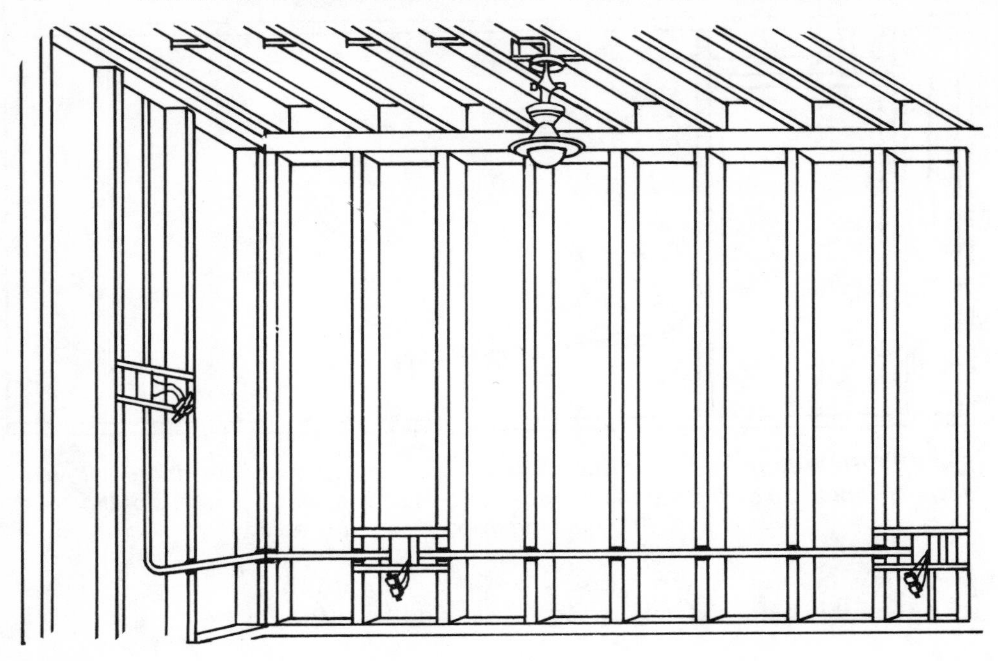

Fig. 7-12. Thin-wall conduit comes in 10-foot lengths and is installed by notching studs.

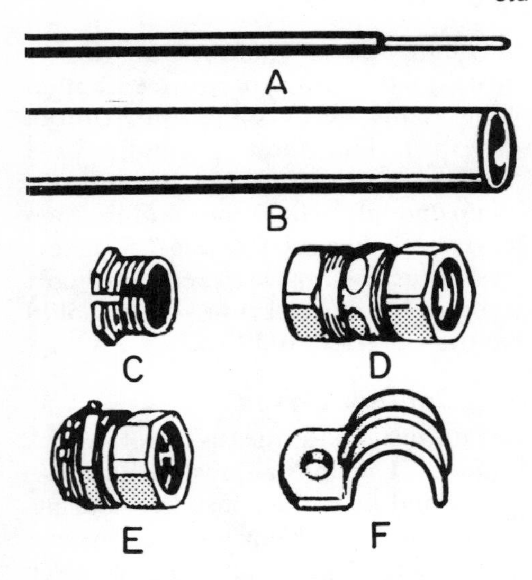

Fig. 7-13. Components for working with thin-wall conduit (electro-metallic tubing or EMT), (A) Type TW color-coded wire that is run through the tubing; (B) the conduit; (C) an adapter for attaching EMT to the rigid conduit, (D) a thin-wall coupler; (E) box connector, and (F) a strap that is used every 10 feet on concealed runs, every 6 feet for exposed runs.

Bends are made in EMT with the hickey conduit bender, using a piece of pipe as a handle (Fig. 7-14). To use the hickey, measure off the spot where the bend should start, then attach the tool on the inside of the mark. Place one foot on the straight section and pull the pipe handle until the degree of desired bend is attained. The NEC forbids any bend more than 90 degrees, and no more than four such bends are allowed in a run from one box to the next.

To cut conduit, use a standard pipe cutter or a hacksaw with 32 teeth to the inch. In either case, all burrs must be removed to prevent damage to the wiring. Ream out the cut insides and taper with a file.

Where there are difficult bends, it is wise to also install a pull box to facilitate getting the wires through. These are installed in the same way as regular boxes, and covered with a solid plate after the wires have been pulled through.

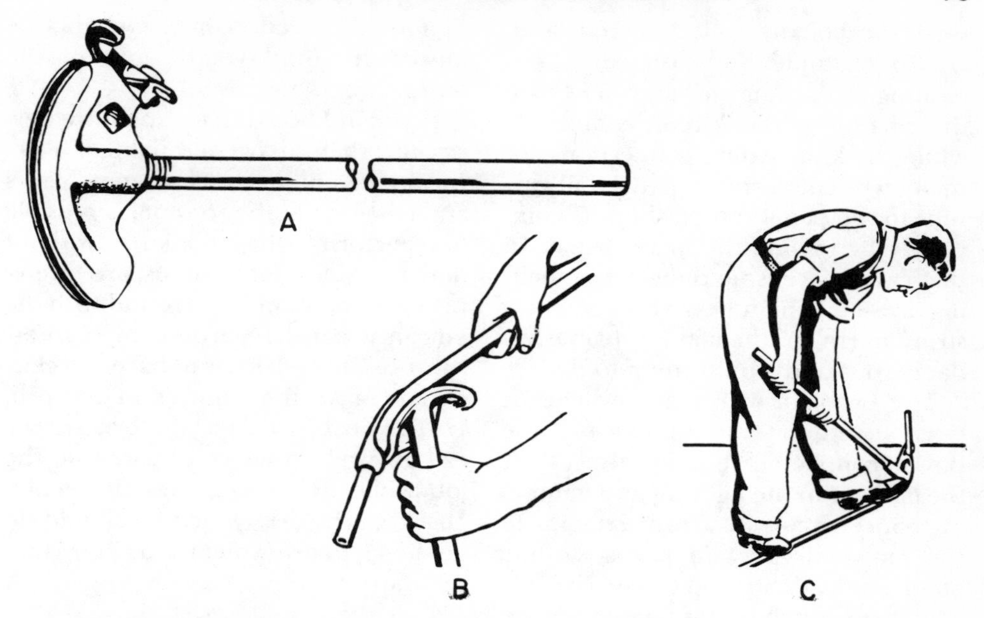

Fig. 7-14. (A) hickey conduit bender with an attached 30-inch pipe handle. (B) insertion of the conduit into the hickey, and (C) technique for bending.

Pulling the Wires Through EMT

After all the conduit is in place, the wires are then pulled through. Wires must be continuous from box to box. Use Type TW wires, and make sure that you purchase the right types to conform with standard color codes. In a two-wire circuit, one wire is black and the other is white. Three-wire circuits have an additional red wire. Beyond that, almost any color can be used except white or green (ground-wire colors). Blue is usually the fourth color, if needed, then yellow, but it is highly unlikely you will need any more than three wires.

Short runs are usually no problem when pulling wires through conduit. They are simply pushed from one box to another. Long runs, especially with several bends, may require the use of fish tape. Run the fish tape through the conduit first, from one box to the next, then bend the new wires around the hook on the tape. It helps to wrap the connection with electrical tape to insure that it won't come undone.

Next pull the fish tape back through the conduit until the wires appear. Cut the wires to size, allowing for connections, then proceed to the next box. When working through bends, work the tape back and forth, using soapstone or similar lubricant, until the bend is passed. Use the aforementioned pull boxes if you foresee a problem.

Making the Hot Connection

Everything that has been discussed so far can be done in complete safety. Since all the boxes, cable, switches, outlets, or whatever have been installed before any connection to the current, there is no way the current can "get" you.

In many cases, you can safely perform the hookup to the current also. If, for example, you are merely extending an existing circuit, you can kill the current to that circuit completely while making your connection. In many types of entrance panels, all current to the panel can be shut off completely by pulling the main switch. In that case, there is no danger of touching live current unless you somehow stray "north" of the main—obviously a dangerous and stupid thing to do.

The problem with main switches is that they don't always cut off the power completely. Some only cut off the current to the lighting and general appliance circuits. Current remains to the range, dryer, and perhaps some other individual appliance circuits. The main switch in this type of service should be labeled main *lighting*, but be mistrustful until you're sure of the setup.

If you are hooking up an entire new circuit, then, have a qualified electrician examine the entrance panel and tell you whether or not it's possible to perform the hookup without danger. Since most of us are pretty nervous about fooling around with the entrance panel regardless of such assurances, my advice is to have the electrician make the connection for you. He'll probably charge you for a service call anyway, so he may as well do the job while he's there. Also have him check your work (which he should do anyhow), to give you an extra measure of security.

Hooking into an Existing Circuit

The last run of cable to the power source will probably have to be fished to the box where the connection will be made. Suggestions for planning this run were given at the beginning of this chapter. Directions for fishing cable through old work are outlined in chapter 8.

To repeat, make sure that the old box is large enough to accommodate two new wires. Study the rules for this on page 85, and make any necessary adjustments. If, for example, the new wiring will exceed the number of connections by only one, and the box contains interior cable clamps, you can remove these clamps, (which count as one connection) and use exterior cable clamps. This may mean cutting away some of the wall to enable you to insert exterior connectors to the existing cable, but this is as easy as any other method. (See Chapter 8 for ways of patching wallboard.)

In some cases, you will have to install a larger box, or gang another box with the old one (see page 86). In either case you will again have to cut away part of the wall. If you are putting in a larger box, remove the existing device and disengage all the old wiring. Remove the old box using a screwdriver or crowbar if it's nailed to the framing. Install the new one as directed on pages 32–35, and reinstall the receptacle, hooking up the old wiring as described previously. (Use masking-tape tags if you have doubt.) Then attach the new wires to the empty terminals, black to copper, white to nickel-colored.

If you decide to replace an outlet with a junction box, simply remove the receptacle and attach the wires together with solderless connectors, white to white, black to black. Cover the box with a solid cap. Never cover up any box with paneling or other material so that it can't be easily located.

Connecting to the Entrance Panel

If you are certain that the main switch cuts off all power to the entrance panel, or if you feel brave enough to mess around in there even if a portion of it is still energized, you can make this connection yourself. Make sure, however, that you know exactly what you are doing, and don't allow your hand to stray beyond the immediate area in which you are working. As recommended previously, most people should hire an electrician to do this phase of the work.

If you do decide to do work on the panel yourself, do it in daylight and come armed with a flashlight. Disconnect the power completely by throwing the main switch. This may be a lever on the side of the panel, a pull-out fuse drawer, or a ganged breaker labeled main. As noted previously, if the breaker is marked main *lighting,* it will disconnect only the power to the general-purpose and general appliance circuits. The range, dryer, and some other individual appliance circuits will remain energized, so extreme caution is advised. If you work only in the panel area not served by these individual appliance circuits, there should be no problem. But don't let your hands wander! Touch only those wires you have to, and use pliers with an insulated handle. Don't touch bare wire with your bare fingers. Use your neon tester to make sure that there is no power where you'll be working. (Put one prong on each power lug or breaker and the other on the neutral bus bar.)

Technique

Plan carefully before you do anything. With the main switch thrown, remove the screws (if any) on the panel cover to expose the wiring. Determine where there is an empty circuit, where your cable will enter the panel, and the path that it will take (Fig. 7-15). Remove the knockouts for the breaker and the cable connection to the outside of the panel. Before cutting the cable, make a trial run to make sure you have enough wire to fit (Fig. 7-15).

Select a breaker to fit the type of panel and size wire used (15-amp for #14 wire, 20-amp for #12). Cut the cable and strip it as you would any other device with terminal screws. Secure cable to panel with standard connectors. Attach the black wire to the breaker and the white wire to the common grounding bar (see Fig. 7-17). Check to see where the bare or green-covered ground wires from the other circuits are attached. In most panels, these will also be attached to the grounding bus bar. If there is a separate bar, connect your ground wire to that. Push the breaker into place as shown (Fig. 7-17).

When everything is in order and all connections are tight, replace the front of the panel and reset the main switch. Set breaker at "on." (Fig. 7-18). Now comes the moment of truth when you can put your handiwork to the test. Check each outlet, switch, and fixture on the circuit to make sure they work right. If everything works, congratulate yourself. If one or two devices don't work turn off the power to the circuit and check out the individual outlets or whatever. The wiring may be loose, incorrectly hooked up, or something similar. But the trouble should be localized there. The circuit is probably okay in general.

When the breaker immediately flips off, you have a different, bigger problem. You'll have to check out the entire circuit to determine the problem. If it's not obvious from a quick check (like a short circuit), I would suggest that you

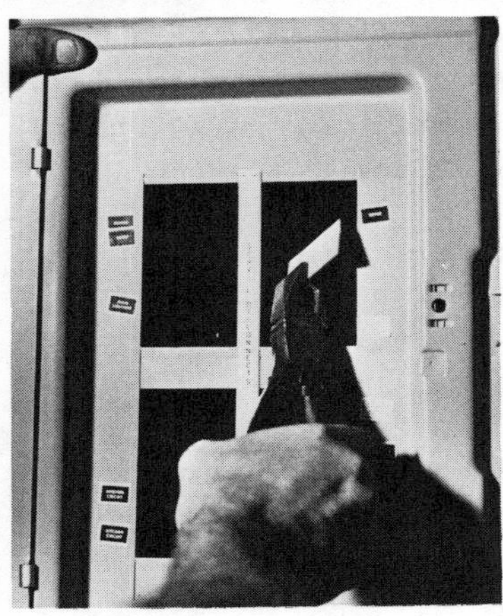

Fig. 7-15. (Left) Find a blank space in the service panel to insert the new circuit. Remove the piece of sheet metal which indicates a circuit not in use. (Above) Run enough cable over to the panel to make connections to both the breaker and the neutral bus bar at the right of this service.

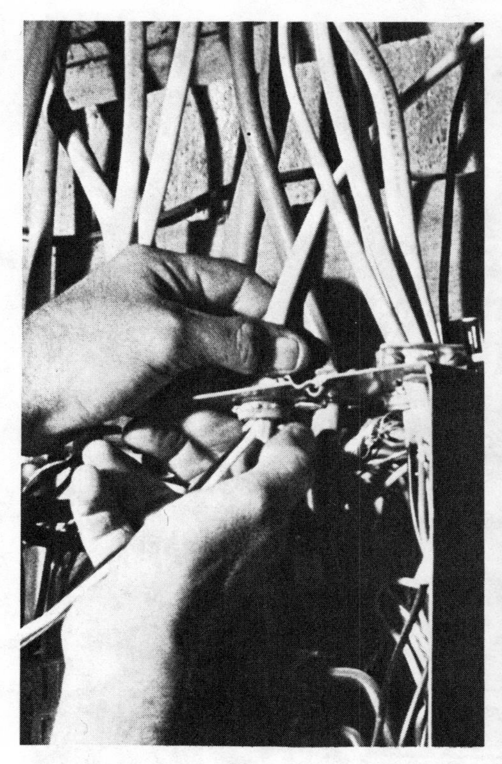

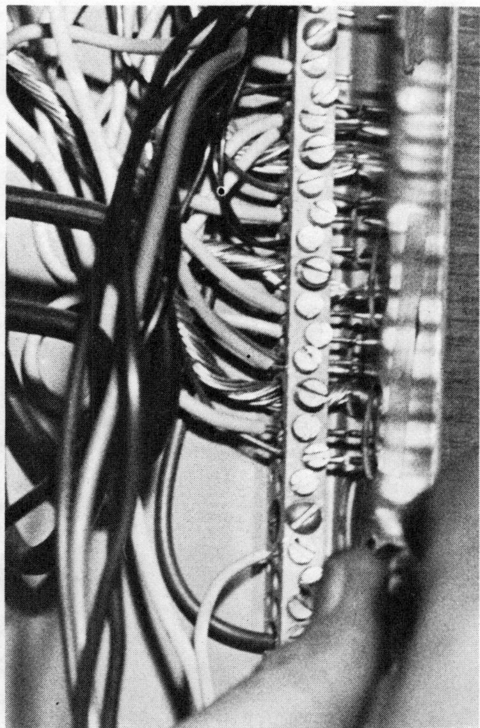

Fig. 7-16. (Left) Use regular connectors to secure the cable to the outside of the panel. (Right) Insert neutral and ground wires into the terminals in the neutral bar and tighten screw over the wires.

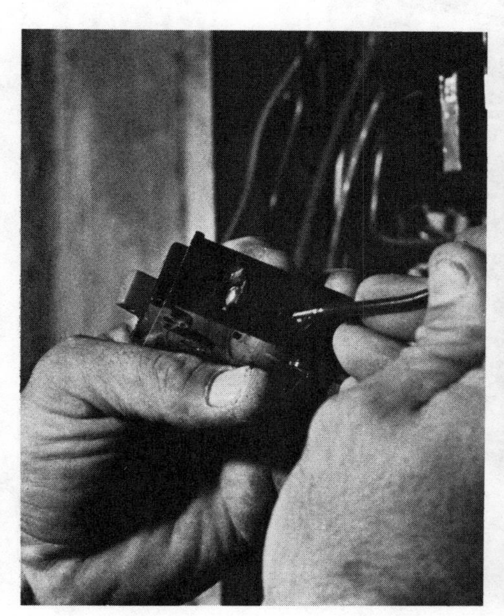

Fig. 7-17. (Left) Attach black wire to the breaker as directed by the manufacturer. This Murray breaker has a standard terminal screw. (Right) Slip the breaker into the slots provided.

call in an electrician and have him diagnose the difficulty.

Fuse boxes are wired similarly, and should be hooked into following the same general directions. Where (and if) an opening exists, screw the black wire to the fuse lug, and the white and ground wires to a common bus bar or wherever the other circuits are attached.

Fig. 7-18. When all is in place, set the breaker to On and pray. If it doesn't trip back to Off, all should be well.

What Size Wire to Use

Number	*Capacity*	*Use*
14	15 amps	General-purpose, lighting circuits (some local codes do not allow this size)
12	20 amps	Kitchen, lower wattage appliances, almost any household circuit
10	30 amps	Subpanel connections, heavy-wattage appliances
8 stranded	40 amps	Service wiring, electric ranges
6 stranded	55 amps	Service wiring

Note: Smaller wires are available, but are for use in low-voltage wiring, lamp cords, appliances, and so on, not for house wiring. There are larger sizes, too, but they are used industrially or beyond the home service. Sizes larger than #2 are designated 1/0, 2/0, 3/0, and so on. In this sequence, the higher the number above the zero, the thicker the wire.

Computing Box Capacity

To prevent dangerous overcrowding inside boxes, the National Electrical Code has set a maximum capacity for each one. The terminology is a little confusing, since capacity is stated according to the number of "conductors." The term is applied to all of the electrical entities inside the box, including the wires, device, interior clamps.

Here's how to "count" them:

"Conductor"	*Number of Conductors Assigned*
Each black and white wire, both leaving and entering	1
Switch, receptacle, or other device	1
Fixture stud, hickey, or built-in clamps (one or more)	1
Grounding wires (one or more)	1

Notice that each interrupted wire, black or white, is counted as one conductor. On a receptacle at the end of a circuit, for example, there would be one black and one white wire, counted as two conductors. In the middle of a run, there would be four such wires, counted as four conductors. All ground wires are counted as one, no matter how many are actually inside the box. The same applies to built-in clamps or fixture hardware inside the box.

Not counted are exterior clamps, pigtails, or fixture wires. Any wires that continue through the box uninterrupted, as may happen in a conduit pull box, are counted as one.

Maximum Number of "Conductors" per Box

Type	*Size*	*Capacity (cu. in.)*	*Conductors Allowed #14*	*#12*	*#10*	*#8*
Rectangular switch or gem	3 x 2 x 2¼	10.5	5	4	4	3
	3 x 2 x 2½	12.5	6	5	5	4
	3 x 2 x 2¾	14.0	7	6	5	4
	3 x 2 x 3½	18.0	9	8	7	6
Square	4 x 1¼	18.0	9	8	7	6
	4 x 1½	21.0	10	9	8	7
	4 x 2⅛	30.3	15	13	12	10
	4 11/16 x 1¼	25.5	12	11	10	8
	4 11/16 x 1½	29.5	14	13	11	9
Round or octagonal	4 x 1¼	12.5	6	5	5	4
	4 x 1½	15.5	7	6	6	5
	4 x 2⅛	21.5	10	9	8	7

If using different sized wiring, allow each #14 wire 2 cu. in., #12, 2.25 cu. in., #10, 2.5 cu. in., and #8, 3 cu. in.

Ganging Boxes

Sometimes it is convenient to have several switches or receptacles together, or to combine them in one central spot. To do this, boxes must be ganged, or installed side-by-side. Most boxes (switch-type, at least) have removable sides that enable two smaller boxes to be combined into one. Simply remove the screws on the sides of the boxes to be ganged, take out the sides, and replace the screws. Only metal boxes can be ganged.

Working with Existing Structures

Electrically speaking, working with existing structures is no different from working with new structures. The same principles apply to both. Adding to or changing the wiring in "old work" is a lot more difficult, however, because the cable is covered by walls and ceilings. Not only is it not visible, but it can be a frustrating and painstaking job getting through the framing and finishing materials without tearing them apart.

In the previous chapter, and throughout the book, "fishing" has been mentioned. If that word conjures up a pleasant vision of lolling by a stream with a line in the water, you are in for an unpleasant surprise. The type of fishing we're talking about here is a difficult, uncertain, and often maddening task. The less you have to do of it, the better.

Drawbacks

When you are doing your own construction, you can run cable wherever you please (within reason, of course, and in conformance to the Code). The framing is there in the open, and you easily figure out the best route from Point A to Point B. You can use any one of the several methods mentioned in the previous chapter. You can choose a direct route or a more circuitous one, if you want to. Furthermore, you know that you've used 16-inch or 24-inch centers for the framing. If you decide later to add another fixture or outlet, you have an excellent idea as to where the studs and joists are.

Eccentric Framing

Not so with old construction. And the older it is, the more in the dark you will be. Homes over 50 years old were custom built. There were no "developments" then, no common plans, and no firmly established framing patterns. There weren't even many building codes.

In many cases, builders and carpenters did use 16-inch centers when framing, but very often they just used common sense. (If they didn't, the house wouldn't still be standing.) Very often, they *over*built. You may well find several studs where one would do. There may be 2x6s or 2x8s in walls. You can find horizontal members for no apparent reason, even diagonal braces where none are needed, or, studs left out if the builder didn't think he needed them.

In any case, unless you are completely renovating a room and don't mind tearing the old walls apart, you will have to try to string cable through an uncertain maze of framing mem-

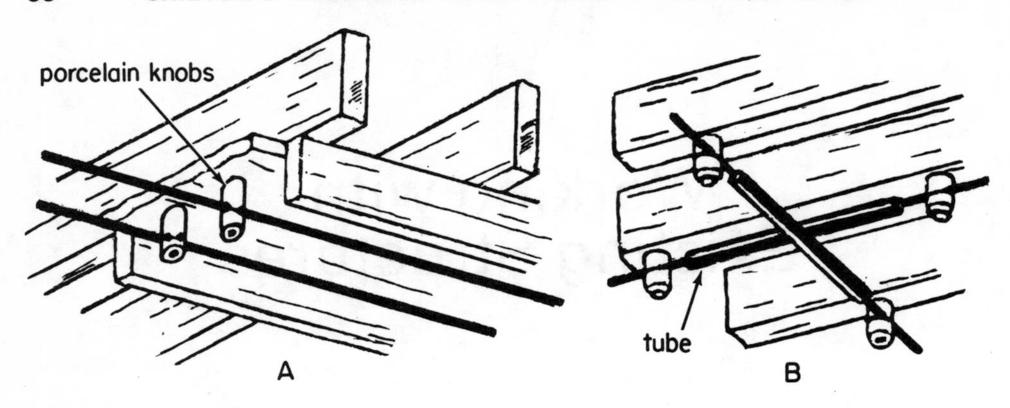

Fig. 8-1. (A) Porcelain knobs kept the wires from touching each other and the framing. (B) Tubes were used where wires crossed and at other potentially dangerous intersections. Such wiring, if present today, should be replaced.

bers, heating ducts, old wiring and God knows what else behind the wall materials. In many cases it would literally be easier to take down the finishing materials to replace the wiring. But who wants to tear down beautiful, strong, old walls of real lath and plaster? Why disturb mellowed oak paneling and woodwork?

Inadequate Housepower

Older homes, however, are surely in need of extra outlets and more housepower. Many need entire new wiring. Truly old homes probably have knob-and-tube wiring (Fig. 8-1), which can be a definite hazard, particularly since its insulation is old and almost worthless.

We would certainly recommend that you *not* try to install new wiring throughout the entire house by yourself. Maybe, after you have done a few smaller fishing tasks, you may want to try this, but not yet, if ever. Call an electrician, and have him give the house a thorough going over. You probably need a new service, new entrance panel, and so on, as well as new wiring. Unless you are very brave and a masochist besides, let the electrician do it all—or most of it.

Assuming, however, that there is enough capacity at the service (a rash assumption unless it's been updated), you can extend an underutilized circuit to include more outlets, put in another switch, or add a new fixture. You can also install a new circuit, if your total housepower allows it.

To determine whether you have excess capacity and to learn how to trace the various circuits, see Chapter 6. Also read (or reread) the previous chapter for suggestions on planning and for general techniques. Most of the same methods and materials are used in existing work that are used for new work. If you haven't read pages 32–35 on installing boxes in old work, do so now.

Running Cable in Old Work

Points of Access

To repeat, the principal difference between working with old and new structures is in running the cable. Sometimes, this is not a big problem. If you are merely adding another outlet or two to an existing circuit, and the

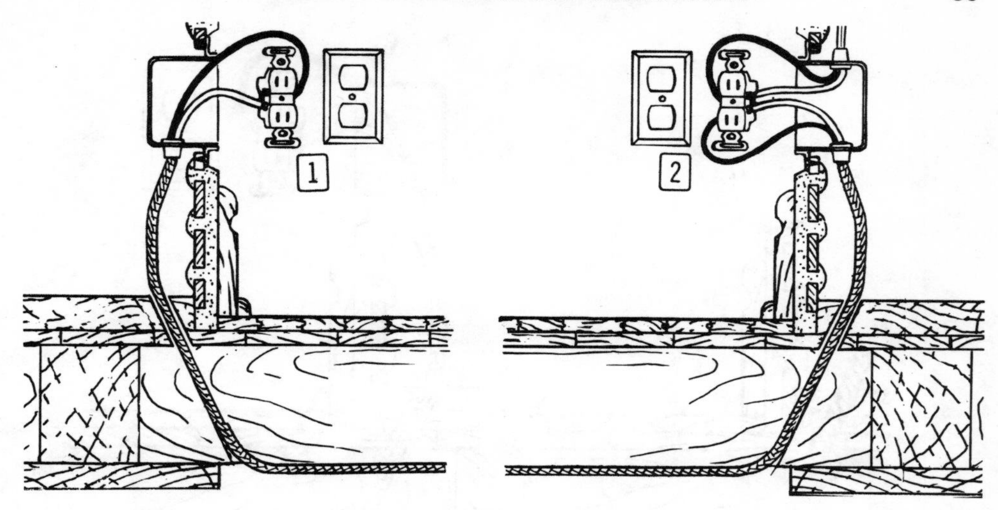

Fig. 8-2. The easiest way to add new outlet (1) is to run new wiring through the basement from former end-of-circuit outlet (2).

floor is accessible from the basement, the cable can be strung without too much difficulty down into the basement and back up again (Fig. 8-2).

Basement

When the outlets are in an interior partition, drill ⅝-inch holes from the basement straight up through the subfloor and bottom plate. On an outer wall, use a long extension bit, and bore diagonally through the plate into the wall cavity as shown in Figure 8-3. Needless to say, make sure that you've lined up the holes with the outlets. And de-energize all circuits in that area, since you may inadvertently drill into some of the existing wiring.

Remove the existing box and device, chiseling it out if necessary, and install the new boxes. (See pages 32–35). Run the cable through the basement as described in Chapter 7. You should now be able to push the cable up from the basement into the wall cavity and the boxes. If you must cut the cable before you do this, leave plenty of slack. You can always cut off any excess later.

With the boxes out, you should be able to grab the cable through the box holes and connect it to the boxes. If you have difficulty, see the section below on fishing. In this instance, a simple bent coathanger may provide all the fishing gear necessary.

Attic

If you have an open attic, you can approach the job from above in much the same way. If an attic floor is in the way, you will have to lift a few floor boards to gain access to the wall plates. Then simply drill holes and push the cable down to the two box openings.

Baseboards

Those are the easy ways, but if they are not feasible, you can remove the baseboard between the two outlets (in the same room or adjacent rooms) and cut a groove in the plaster or wallboard a few inches above the floor (Fig. 8-4). Check local codes on this first, though. The groove must be deep enough to accommodate the cable without crushing it. It is a good idea to use armored cable or conduit for such a run, so that it will withstand crushing and is less likely to be damaged by errant nails

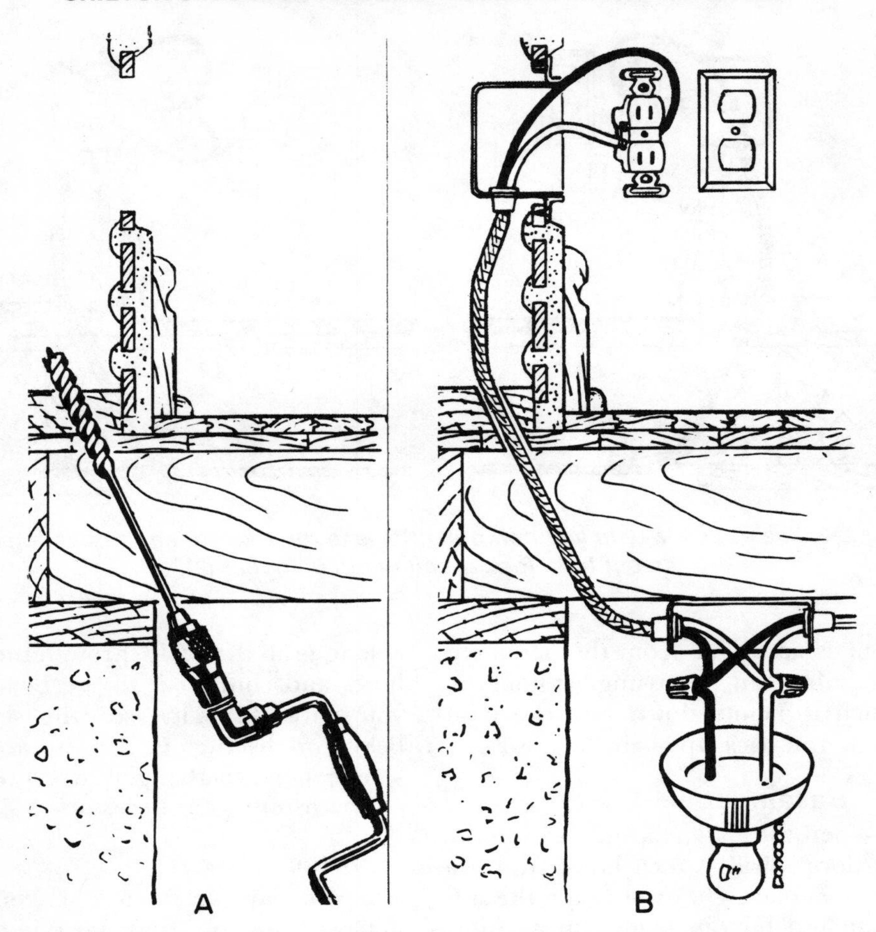

Fig. 8-3. (A) To run cable up to an exterior wall, use a long extension bit to drill up through the bottom plate. (B) Run cable up through the plate to the outlet from an existing power source.

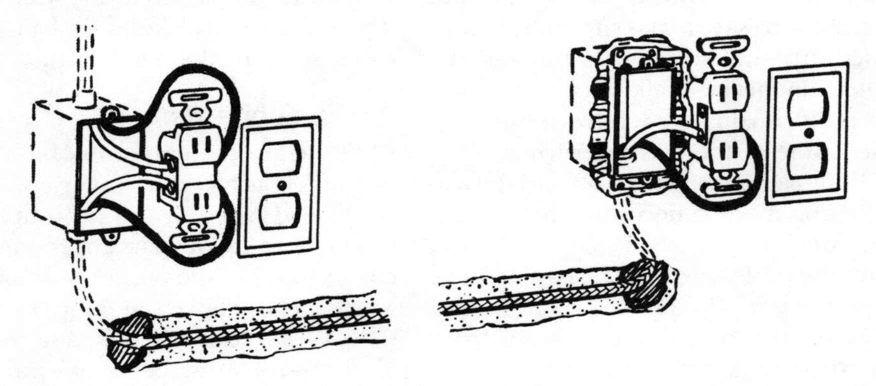

Fig. 8-4. If allowed by local codes, cable can be run behind baseboards to a new outlet. Remove the baseboard and drill holes below old and new outlets. Chisel out a groove deep enough to hold the cable in the plaster between each hole. Be careful when replacing the baseboard that you don't nail through the cable.

when you replace the baseboard later. If you use Type NM cable, cover it with thin steel plates at the studs (see p. 75). Use care when reinstalling the baseboard, nailing well above and below the cable.

Doorways

If your run includes a doorway, remove the trim molding. Notch and groove the framing as necessary for the cable (Fig. 8-5). Nonmetallic cable will almost certainly have to be used here because of the relatively sharp bends around the door frame. Again, check the local code. If the outlet is not too far from the door, you may be able to drill through the door-frame studs to the box opening, using a long extension bit.

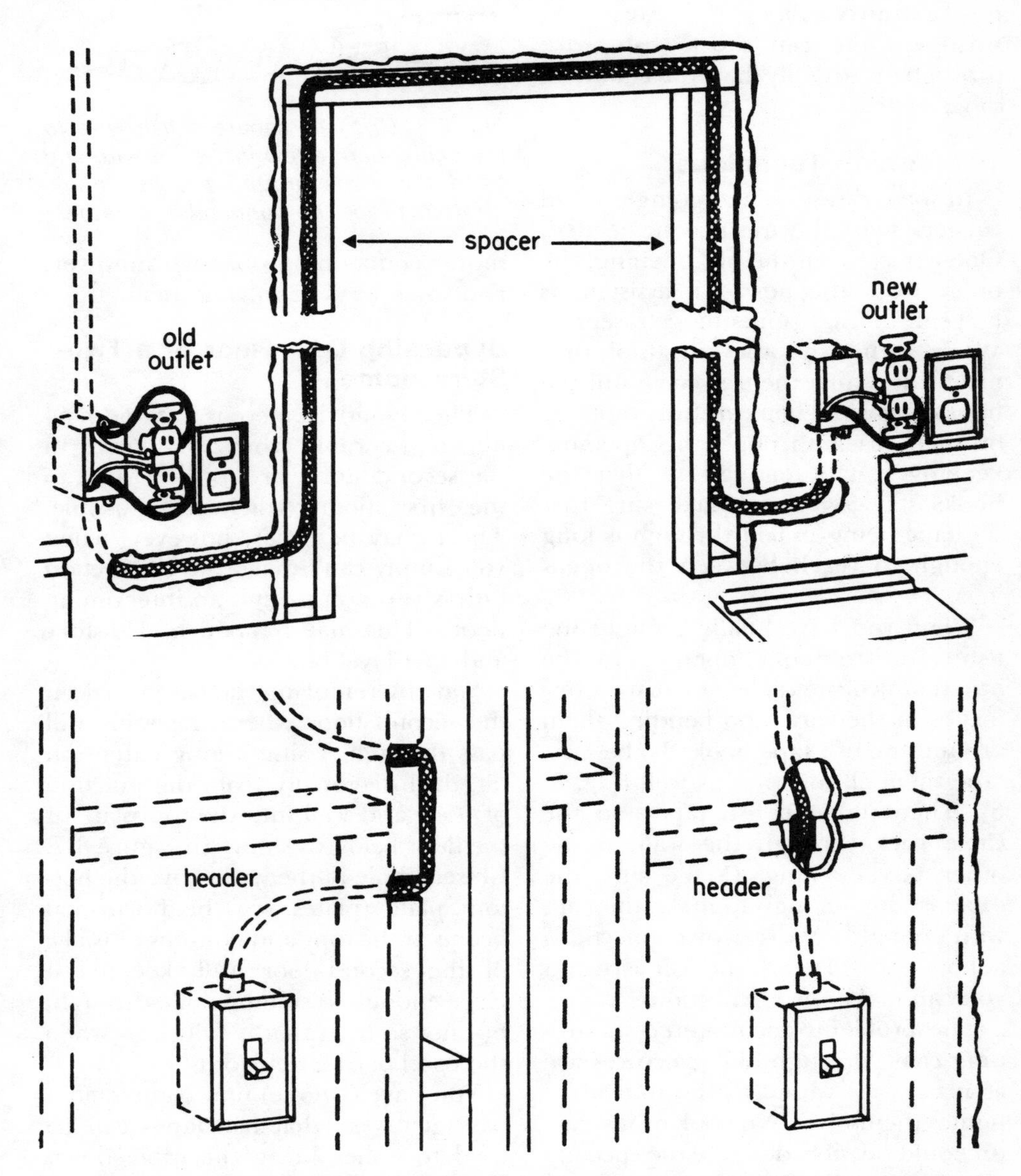

Fig. 8-5. Cable run around a doorway or window opening (top). When headers are in the way, go around them (bottom left) or notch them (bottom right).

After running the cable, replace baseboards, moldings, and the like. Some patching of plaster or wallboard may be required, but this should be minimal. See the end of this chapter for tips on patching.

The Art of Fishing

For all but the shortest runs, some fishing will probably be required. Normally, the fish wire is worked into the new box opening and back to the power source. Since it is steel and springy, you can work fish wire through more easily than you could the cable itself.

Fig. 8-6. (Top) To secure a fish wire to the cable, bare a few inches of wire and bend the wires around the fish hook. (Bottom) Tape the connection together.

Two-Person Technique

In some cases, such as going around corners, two fish wires may be needed. One comes from the old opening, the other from the new. An assistant is helpful here, sometimes essential. Work both fish tapes in until they meet, then turn them slowly until the hooks engage. Then carefully pull one of them back through the opening, keeping it taut enough so that the hooks stay together. Make sure that the tape being pulled through is long enough to reach between the openings.

When you have finally brought the fish wire from one opening to the other, attach the cable by baring a few inches of the wire, and bending them around the fish tape hook. To be safe, tape them all together as well (Fig. 8-6). Then pull the fish tape and the cable back through the wall to the other box opening. (Make sure the cable is long enough to make the run, with enough left over for connections at both ends.) Detach the fish wire and you can make the connections.

The problems encountered in running cable through old structures are as many and varied as the individual homes themselves. No book or instructor could possibly describe the specifics for getting cable through every type of obstruction. Here are a few of the more common problems, however, and some ways of solving them.

Bypassing One Floor of a Two-Story Home

The best advice here is to avoid having to run cable from a basement to the second story, or from the attic to the first floor, whenever possible. There may be times, however, when you simply can't make the connection unless you go through an intervening floor. (This may happen in 1½-story and split-level homes.)

Start by removing the baseboard on the second floor where the cable will pass through. Using a long extension bit, drill diagonally from the junction of floor and wall into the top plate of the floor below as shown in Figure 8-7. Chisel a hole in the wall above the bottom plate (plates are the horizontal beams at the tops and bottoms of walls) of the second-floor wall, keeping it large enough to run the cable through, but not so high that it will show when the baseboard is replaced.

You have (I hope) now completed a passageway so that fish tapes can be used to either bring the cable down from the attic, or up from the basement, as the case may be. Two fishers

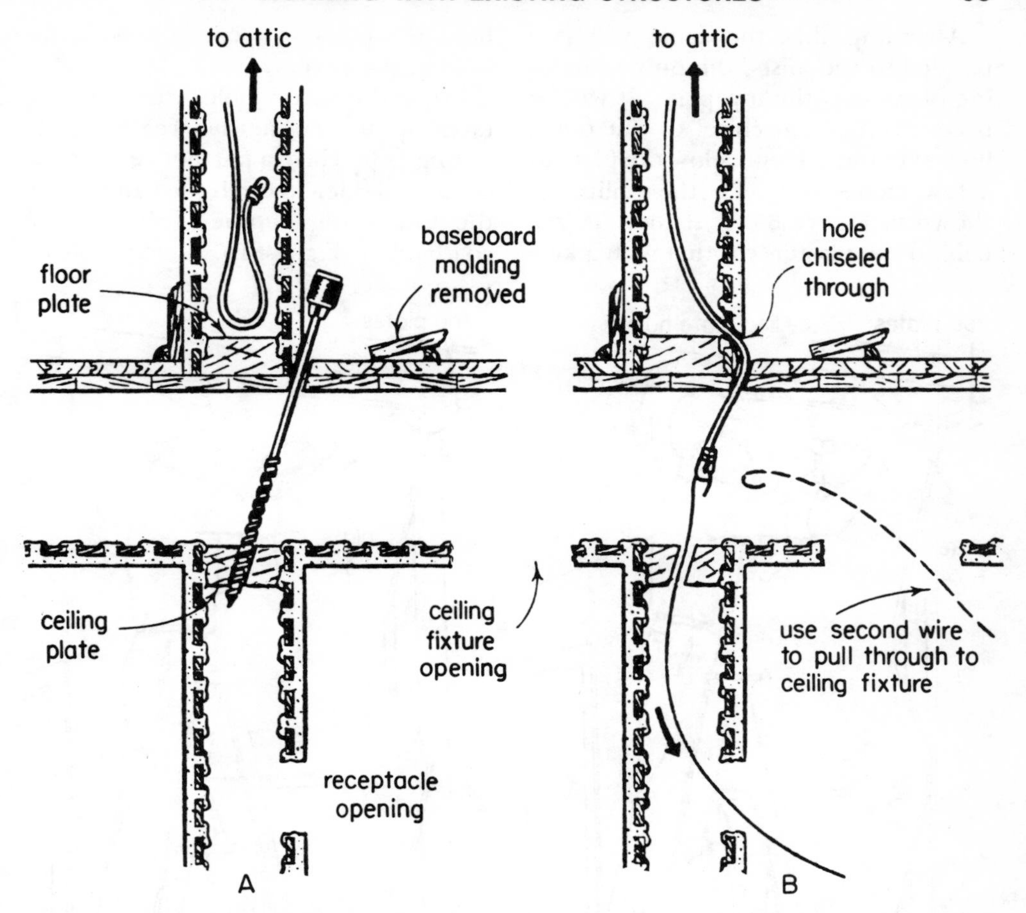

Fig. 8-7. When cable must be strung through an intervening floor of a two-story home, (A) remove the baseboard from the second story and drill down diagonally through the top plate of the first story. (B) Chisel a hole above the bottom plate of the second-story wall to get the cable around the plate then fish the cable down to the outlet below. If you are adding a ceiling fixture on the first floor, a second fish tape will be needed to get the cable over. Similar techniques are used to get cable up to a second story from the basement.

will be needed here. One person runs the tape up or down, the other grabs it (by hand or with another tape) and brings it around the second floor bottom plate and through the hole in the first floor top plate.

Getting to a Ceiling Fixture from Wall Below

When the cable must run from a wall switch or outlet to the ceiling above, the simplest method is to run the cable up to the attic and over to the fixture. However, if you don't have an attic or if that part of the attic is inaccessible, you have another problem.

The first thing to do is to plan the wiring so that the cable will run parallel to the ceiling joists. That way you can fish from the wall with only the top plate of the wall as an obstruction. If you try to run across the joists, you will have to go through all of the joists in between. Avoid this at all costs, but when all else fails, see below for a difficult but workable solution.

Assuming that the cable will run parallel to the joists, the only remaining obstacle is the top plate. It will be necessary here to chisel or cut out a hole extending from below the plate to a few inches out onto the ceiling as shown in Figure 8-8A. If there is lath behind the plaster, cut that with a keyhole or saber saw to conform with the hole in the plaster.

From the access hole, run one fish tape up to the newly created wall-ceiling hole. Have a partner ready on a chair or ladder to catch the fish tape at the top, using another fish tape or coathanger (Fig. 8-8B). Attach cable to

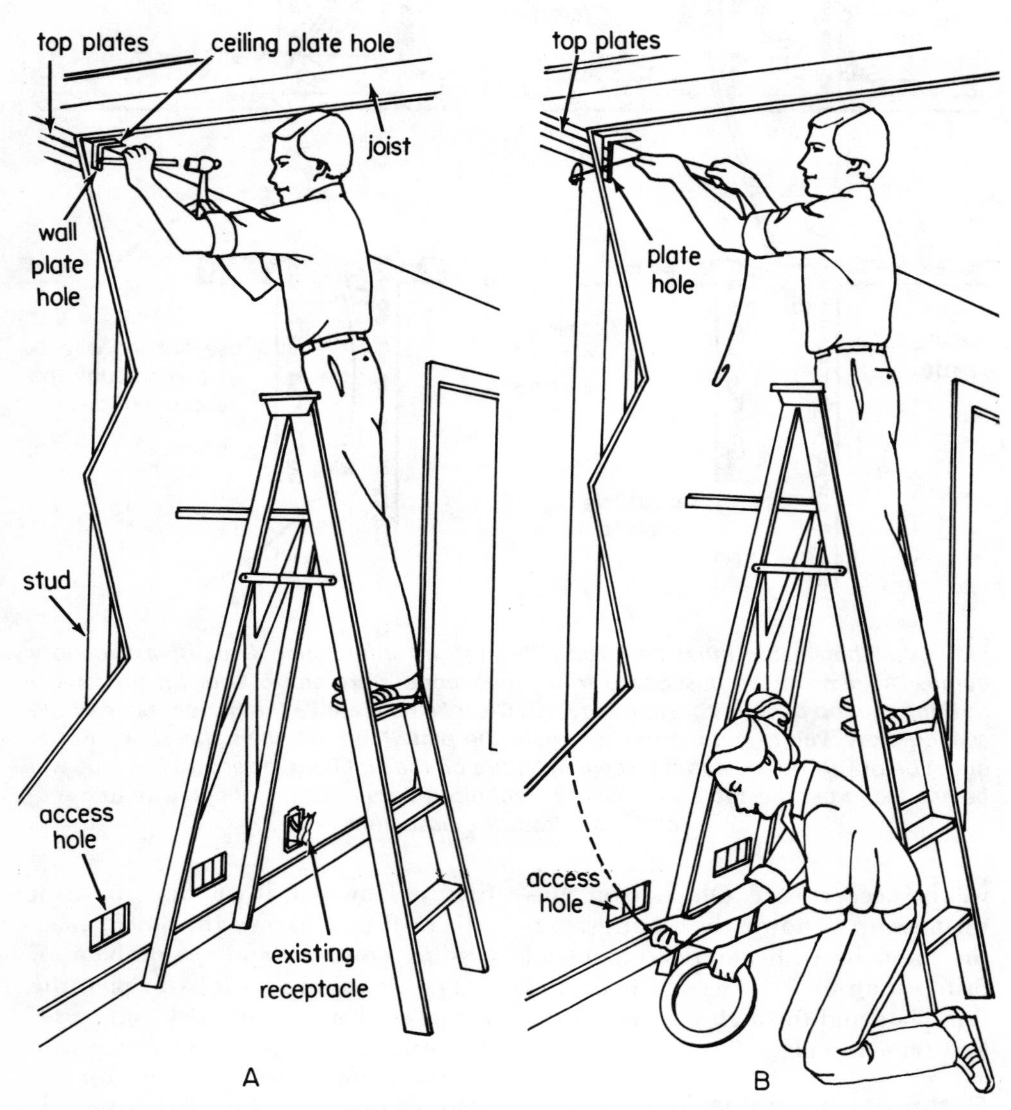

Fig. 8-8. (A) The first step in getting from an existing outlet to a ceiling receptacle not in the same line is to first cut access holes from the outlet to a hole below the joists where the ceiling box will be located. Another access hole should be chiseled out at the wall-ceiling juncture as shown and described in the text. (B) One person runs fish tape up to the wall-ceiling hole from the access hole below. Another catches the tape through the plate hole at top.

the end of the first fish tape and pull it back down to the access hole in the wall (or basement). Now have your partner move to the precut hole for the fixture in the ceiling. Push the fish tape through the wall-ceiling hole toward the hole. Have your partner grasp with another fish tape, pull that tape back through the corner hole, and attach the cable as previously described. Your partner then pulls the cable through to the ceiling box (Fig. 8-9).

If the wall and ceiling material is less than ½ inch thick (most is), cut a groove in the plate to a depth of ½ inch and insert the cable into that (Fig. 8-10). Drive a staple over the cable to hold it. For finishing materials more than ½ inch thick, simply staple the cable to the plate and patch the hole as described at the end of this chapter.

Working Across the Framing

Of all the methods of stringing cable in old construction, this is the least desirable. Use it only when no other

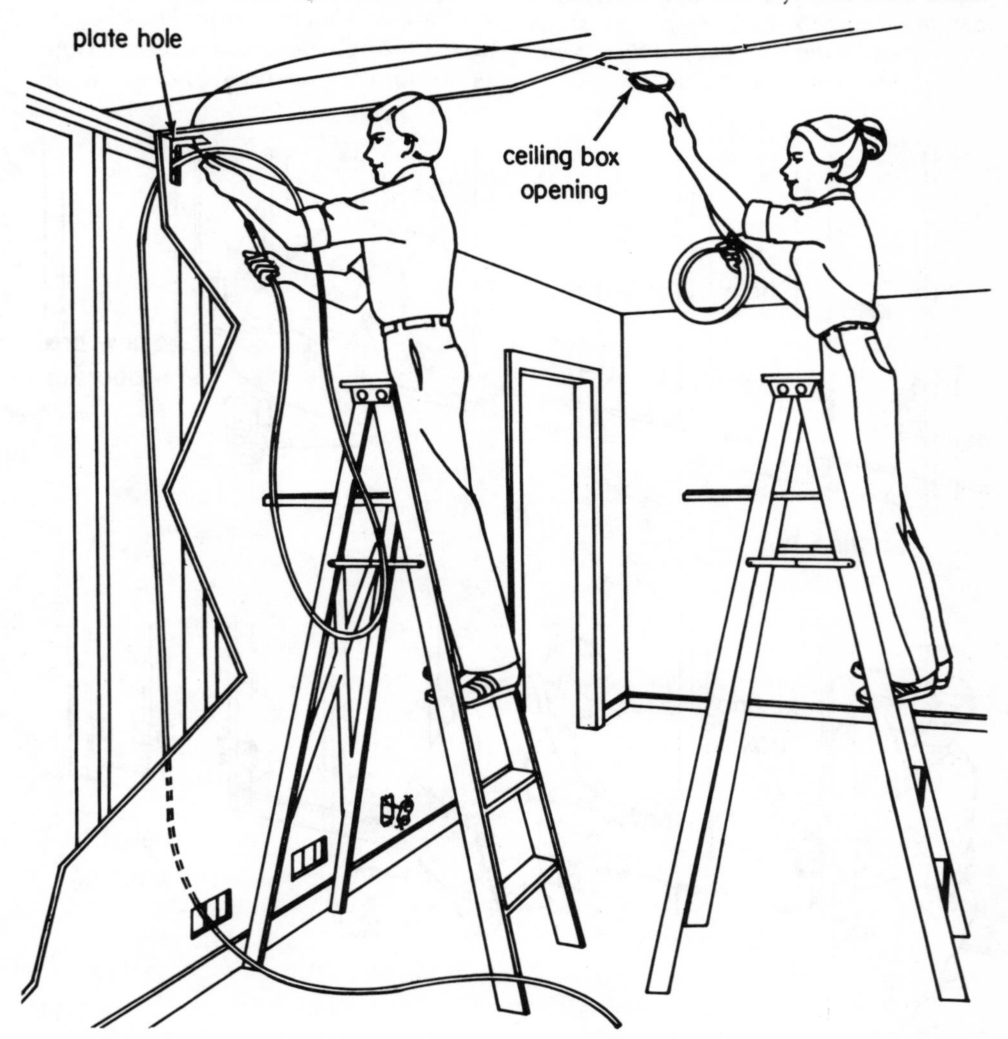

Fig. 8-9. A fish tape is run from the ceiling-box hole to the plate hole. A second person then attaches the fish tape to the cable, and the first person pulls it back through to the ceiling box.

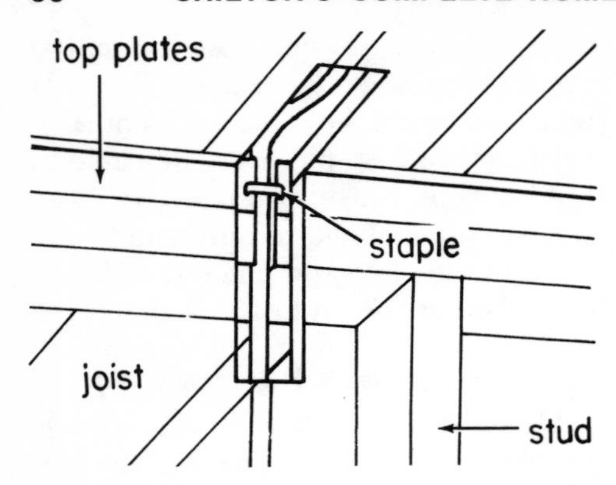

Fig. 8-10. If the wall finishing material is less than ½ inch thick, notch the studs and plate to a depth of ½ inch and staple the cable into the notch.

method is possible. Those, for example, with homes built on a slab and with little or no attic room, may have no choice. Figure 8-11 shows a typical route across studs, but the same principles are easily transferable to joists.

First cut out the hole for the new box (Fig. 8-11A). Either remove the existing box altogether or the knockout in the direction where the cable will travel. On the same level as the boxes, locate studs by tapping (see page 34) and cut access holes at each stud about 4 inches wide. Drill ⅝-inch holes through each stud, keeping them as straight as possible under the cir-

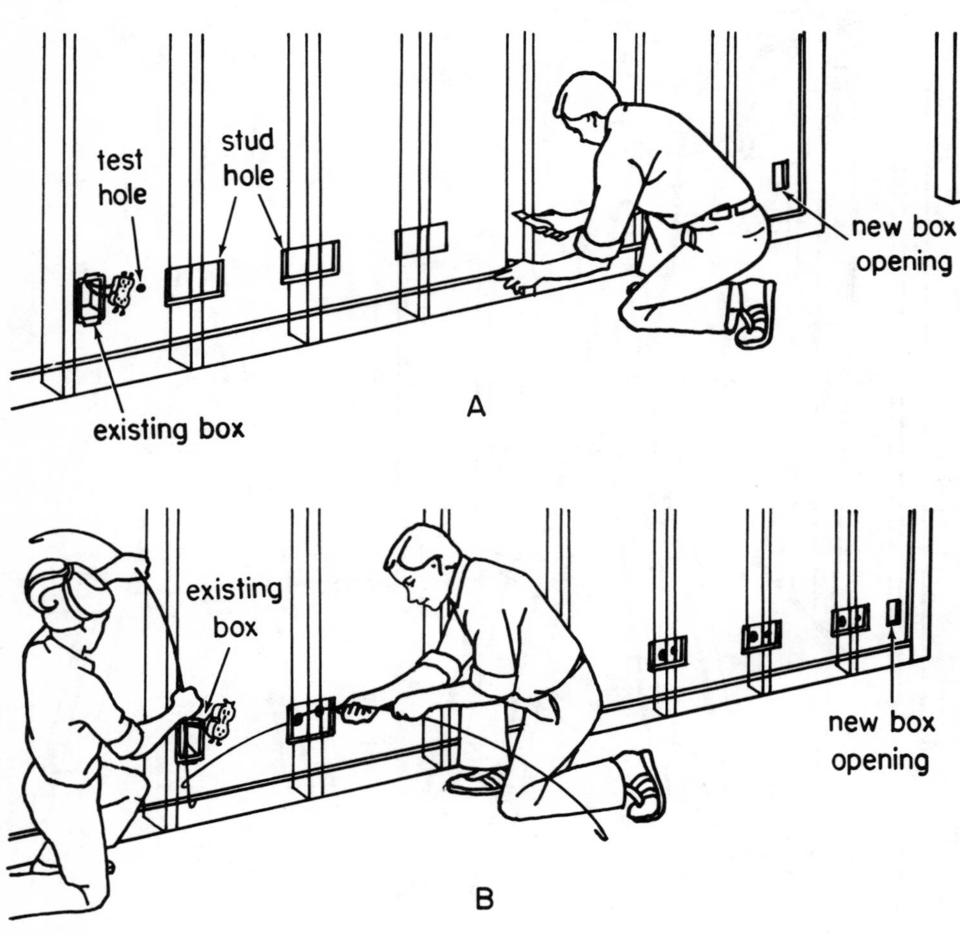

Fig. 8-11. (A) Cut access holes so that you can drill through studs to get cable across a wall. (B) Run the cable from an existing box to a new one through access holes, using a fish tape.

cumstances. Right-angled drilling attachments are available.

Here again, you will need a helper and two fish tapes. Start with the existing box and run the tape to the nearest stud (Fig. 8-11B). Your partner should be ready with another tape pushed through the opposite side of the hole in the stud. Let him grasp the end of your tape. Pull this back through the stud hole to your end and attach the cable. Now your partner should pull the other fish tape back through that stud. Remove the tapes and do the same thing at the next stud. Repeat until you reach the next box.

Depending on the circumstances, it may be easier to have your coworker start from the new box and run his tape all the way through the existing box. Either way, this is not easy, and should be used only as a last resort. Added to the misery is the fact that all the access holes for drilling will have to be patched up as described below.

Wall and Ceiling Repair

Wall repair depends on how the wall was constructed to begin with. Ceilings are replastered in much the same way as walls, except that there may be further problems because of the law of gravity.

Common Materials

There are three common materials used for wall and ceiling repair. Although each has its own special characteristics, they are all basically made of powdered gypsum. In most cases, they can be used interchangeably. However, if you are contemplating extensive renovations, you should know the differences.

Spackling Compound

This consists mainly of gypsum powder and is available either in dry form for mixing or in cans already premixed and ready to use. The dry powder is cheaper, but for small jobs the premix is easier to work with. Use spackling compound to repair small holes in plaster and wallboard.

Joint Cement

This is smoother and thinner than spackling compound and is used for closing seams between wallboard sections. It can be neatly feathered out (spread to a very thin, blending edge) to fill any gap without leaving telltale borders. Both spackling and joint compound can be sanded, painted, and papered over without showing any traces of the previous defects.

Patching Plaster

This is similar to spackling compound but contains fibers that make it effective in filling large gaps. It is highly cohesive and doesn't shrink, but feathering it out is tricky work. Use it for larger holes.

NOTE: A few gypsum products, including some forms of spackling compound and joint cement, contain asbestos fibers. The incidence of lung cancer is 16 times higher among asbestos workers than the general population. This may be a result of prolonged exposure, but the families of asbestos workers have been known to suffer from the disease, presumably merely because the fibers were brought into the house. Avoid any wallboard-patching product that contains asbestos. Most are now being made without asbestos, but some asbestos compounds are still on the market. Read labels carefully.

Patching

There are several ways to patch a large hole in wallboard, but you cannot

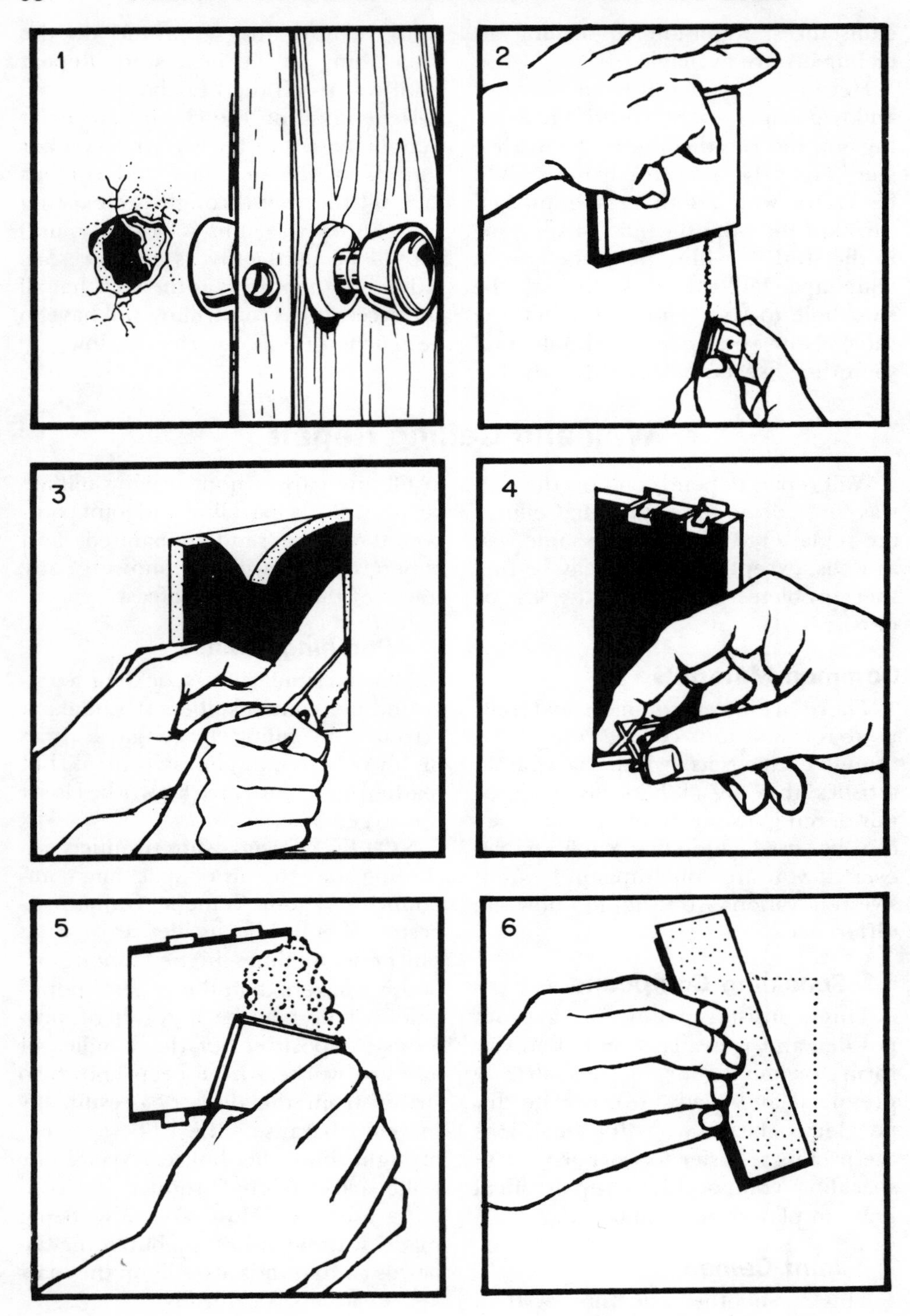

Fig. 8-12. A drywall-repair kit with plastic clips. (1) A typical "doorknob hole" (2) fit patch to hole (3) cut hole (4) apply plastic clips (5) apply spackling compound to patch (6) smooth compound.

do it simply by pressing spackling compound into the hole, for there isn't anything back there to hold the spackle in place.

One method makes use of a scrap of wire mesh, held in place with string, to provide the necessary backing. Another method calls for cutting a triangular wallboard patch with beveled edges that grip the existing wall. A third, much simpler, method makes use of a set of specially designed plastic clips to hold a patch of any desired shape in place while it is being spackled (Fig. 8-12). Clips are available separately for larger holes, but you're more likely to find them in a patching kit, which contains all necessary tools in disposable plastic form. Included in the kit is a small amount of spackling compound and a 4x4-inch patch, which is just the right size for "fish-holes," covering gaps left by removed boxes or holes made by banging doorknobs, some common jobs.

For smaller holes, particularly down near the baseboard, stuffing the back of the hole with newspaper is sometimes effective.

When a large section of wallboard is damaged, it is often best to replace the area entirely. You don't need to remove the whole panel, just cut completely around the damaged area with a keyhole saw; then extend the cuts horizontally until you reach the center of the studs on either side. Make your cuts carefully so you don't damage concealed wiring or plumbing, and work with a utility knife in the vicinity of studs so you don't saw the lumber. Then use the claw of a hammer to clean away old nails and gypsum residue.

Fig. 8-13. After the new wallboard section is nailed to the studs, the joints are filled with joint compound.

Using scrap pieces of gypsum wallboard fill the cut-out space. Nail the exposed half of the studs and fill the cracks with joint compound (Fig. 8-13).

High and Low Voltage

In most homes built since World War II there are three wires leading from the service to the entrance panel. Two are hot wires that carry 120 volts each, the third is the neutral wire maintained at zero voltage. Acquiring higher voltage is a simple thing for an electrician to do. All he does is bring in a second black wire to the service. The 240 volts exists between the two black wires; between either black wire and the white neutral wire is the usual 120 volts.

When the power source is distributed through the entrance panel, only one of the incoming power lines is used for most circuits. Certain home installations, however, need both of these 120 lines to supply the necessary wattage. Even a 30-amp circuit, for example, can deliver only 3,600 watts with 120 volts, not enough for a typical range rated at 8,000 to 16,000 watts. If you live in an area where natural gas is piped in, you may never need 240 volts at one time. Where available, gas is usually used to power water heaters, ranges, and dryers, the most common high-wattage appliances. And don't forget that "split" ranges, where the oven is in one place and the burners in another (such as a countertop) all use 240-volt wiring.

When there are no natural gas pipelines, electricity is used to power these home conveniences. In some areas, a liquified petroleum gas (LPG), such as propane, may be used. Heating oil can also power some types of water heaters, but not ranges or dryers. Two circuits of 120 volts are "tied" together at the service to provide the necessary 240 volts.

If you are leery of tackling 120 volts, be doubly afraid of 240 volts. You may stand a chance of surviving a 120-volt jolt, but 240 volts is more than any body can handle. In this book, I am not going to discuss 240-volt wiring. If you wish to tackle this type of wiring, there are several other good books on the subject. But be warned. It's more complicated, and much more dangerous.

How to Recognize High-Voltage Wiring

You should, however, know a few things about high-voltage wiring. First, you should know how to recognize it.

One of the signs of 240-volt usage is a large three-prong outlet, which is used for portable appliances such as electric dryers. Three-prong outlets come in different styles and shapes as shown in Figure 9-1, but they have one common feature: they are big, and they have a configuration different from that of standard 120-volt receptacles. Each has a different rating and is designed to accommodate a specific plug in only one way. For example, if you try to plug a 50-amp appliance into a 30-amp outlet, you will find that the bottom prong cannot fit into the right-angle prong of the 30-amp outlet. And you cannot plug it in any other way except the right way. However, it is possible to plug an appliance of a given current demand and voltage into an outlet rated the same voltage but a higher current (e.g., a 240-V, 30-amp appliance into a 240-V, 50-amp outlet).

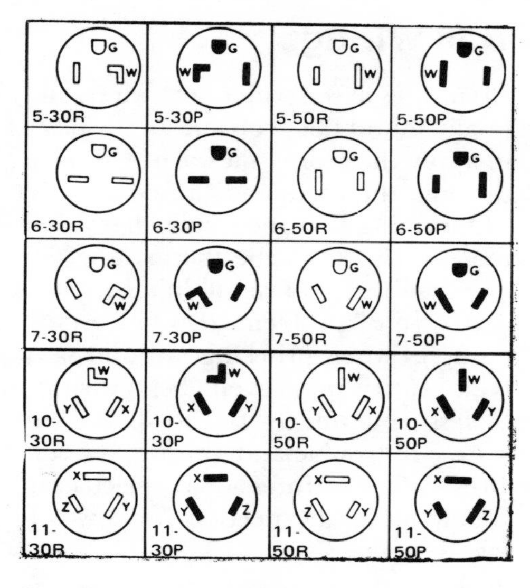

Fig. 9-1. Typical configurations of high-voltage plugs and receptacles. The code after each identifying number and hyphen lists amperes (30 and 50) and type of outlet (receptacle [R] or plug [P]). Receptacles are in white, plugs in black, and they are built so that only the specific plug will fit into its matched receptacle.

Another sign of a 240-line is the wiring itself. If it is visible, as it is for some ranges when you pull them away from the wall, it is much thicker and heavier than the other wiring you're used to. It is probably #6 or #8 wiring, often the same type of cable used for the service entrance and labeled Type SE. The minimum-sized wire that can be used, and only for some appliances such as a separate cooking unit, is #10.

A sure sign of a 240-volt circuit is a ganged breaker in the entrance panel. In a fuse box, you will ordinarily find a pull-out cartridge fuse.

Some Things You Can Do Safely

Because we advised you not to mess with 240-volt wiring doesn't mean that you have to shy away from the appliances served by this type of cable. Once you unplug a dryer or disconnect a range circuit at the service, you can work on the appliance itself in perfect safety.

You can replace the pigtail cord on a dryer, for example. You can put in new burners or heating elements in a range, and save quite a bit in service charges. (Buy them at the company service center that makes the appliance or at any store that specializes in parts for appliance repair and maintenance.)

As long as the equipment is safely grounded, as explained below, you can move a range or dryer to sweep under it without being worried. It is, however, a good safety precaution to unplug the appliance, if possible. And always check any exterior ground wires afterward to make sure that they have not been disconnected.

Appliance Grounding

Because of their high amperage, major appliances must have their cabinets grounded against current leakage. If your home is reasonably new, with grounded three-wire cable, your appliance is grounded through the three-prong plug. Otherwise, a separate ground wire must run from a screw on the cabinet to a cold-water pipe, preferably copper.

Low-Voltage Bell Wiring

Every home has some low-voltage wiring. Since it is unobtrusive and seldom a problem, many of us are unaware of it. Low-voltage wiring is usually #18 or smaller and is often called "bell" wiring because its principal use is for doorbells or chimes. It is also used, however, for heating-unit thermostats, intercoms, and other special applications.

All low-voltage wiring is routed through a transformer, which steps down the voltage drastically—to anywhere from 6 to 18 volts, only a small fraction of the voltage in normal circuits. It is possible to get a minor tingling sensation from bell wiring, but you can't get a significant shock, much less do any harm to yourself. Low-voltage wiring can be worked on without deenergizing the circuit.

Here, then, is one area where even the most craven of us can do our thing in perfect peace of mind. This doesn't mean, though, that tracing the source of a problem is any easier. In fact, it can be more difficult, because bell wiring can be run anywhere.

Let me give you an example. One fall, we had a chilly autumn night, a common occurrence in Buffalo, where I used to live. Lacking a fireplace and with children in the house (and no energy crisis), I tried to turn the furnace on. No response.

I had the furnace (gas) checked out. No problem. I assumed that the thermostat was defective and bought a new one. Still no heat. The problem had to be in the wiring, but where and why? I followed it as best I could and found no loose connections or similar difficulty. The wiring ran under some new tiles I had put up on the basement ceiling. To check that out, I removed all the tiles in the path of the wires, and finally found the problem. In my haste to apply the ceiling tiles, I had stapled through the wires and caused a short.

Assuming that the reader has a few more brains than the then-young do-it-yourselfer, you shouldn't run into that type of problem. (Do be careful, though, when working around any type of wiring, even bell wiring.) Unless you utterly fail to locate the source of trouble and give up, the services of a professional electrician should never be needed for low-voltage systems.

Doorbells

The most common problem encountered with low-voltage wiring is a doorbell that doesn't work. The first step is locating the origin of the trouble, and the most likely place to look is the button at the door (Fig. 9-2). If you have a button at both the back and front door, with one working and the other not, you can be pretty sure that the nonfunctioning button is defective.

Many of us simply hang an "out-of-order" sign on the door and forget it. Let the caller knock. It's good exercise. While sign-hanging may be a necessary stopgap, repair is almost easier than putting up the sign.

Technique

To make sure that the button is the culprit, remove the unit by taking out the screws or by prying it out from the wall if it's a press-in type. There should be two wires connected to the back. Disconnect these from the button and press them together. If bell or chimes ring, the button is defective. Get a new button (most hardware stores carry them) and replace by attaching the two

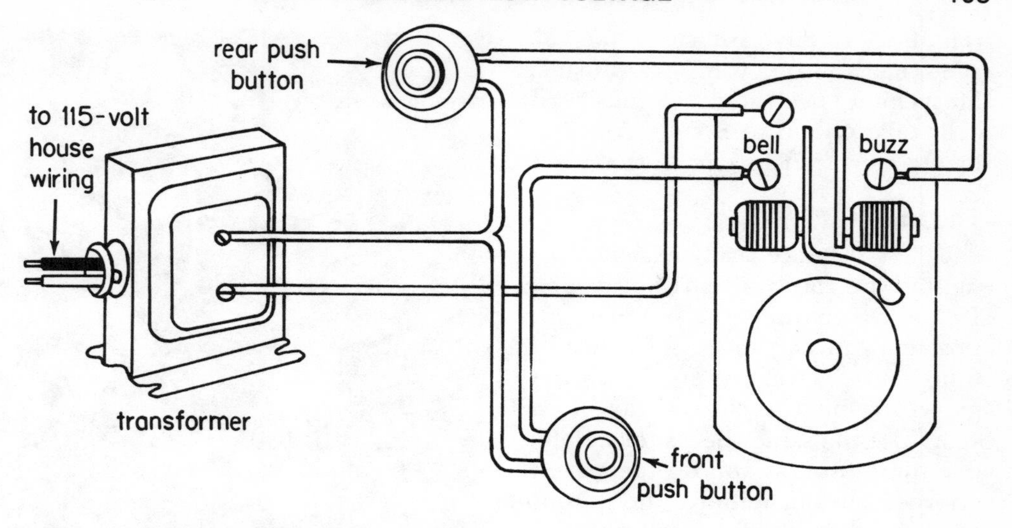

Fig. 9-2. Schematic representation of a typical doorbell circuit.

wires to the terminals. If the button is sunken into the woodwork, try to get a similar kind for replacement. Otherwise, any style will do.

Sometimes the bell or chimes themselves are worn out, corroded, or just plain dirty. If the wire-touching test doesn't ring the bell, suspect such a cause, or, perhaps, a loose connection at the bell. Lift, snap, or unscrew the cover from the unit and look inside, examining the wires carefully where they fit under the terminal screws. Give them a little tug. Remember that the low voltage can't hurt you. If they are loose, tighten the screws and try the button again.

If there is a lot of dirt or corrosion, unscrew the wires and scrape them with a knife. Vacuum or brush the other parts of the unit while you're at it. Clean off the points of bells and buzzers, or remove and clean the plungers that are supposed to strike chimes. (They're activated magnetically.) Use lighter or cleaning fluid or some other form of naptha. Anything heavier, like oil, will clog the plungers further.

After replacing the wires on the terminals, try the button(s) again. Chances are that all will be copacetic. If not, there may be trouble in the transformer or wiring. See p. 104 for what to do about that.

You may decide, whether the unit is defective or not, to replace it with something newer and more attractive. Some of the older bell/buzzer types are rather ugly, and new chimes can be a positive decorative feature, as well as yield a much more pleasant sound

Fig. 9-3. The newer doorbell chimes not only sound better than the bell/buzzer types, they look better too. (Courtesy of *NuTone*)

(Fig. 9-3). In any case, check the voltage capacity on your transformer. Most chimes need about 18 volts, while bells take only 6 to 8 volts. You may need a new transformer in that case.

Thermostats

Unless they are really ancient, thermostats are rarely a source of trouble. If and when they go, however, the problem can be serious. Like heating units and electric power, they always go wrong on the coldest day of the year. Use the same checks for a defective thermostat as you do for a faulty doorbell button. Remove the thermostat from the wall, disengage the wires, and touch them together. You should hear the furnace click on. If so, the thermostat is the problem.

It is difficult to replace a specific part of the thermostat. Furthermore, if the coil, mercury switch, or other part is defective, you would be wise to replace the entire unit. There is yet another factor to consider. At this point, I would like to digress for a moment. I, for one, consider the energy shortage to be very real, and I sincerely believe that it is vital for all of us to conserve as much energy as we possibly can. (See Chapter 12.)

To that point—if you must replace your thermostat—and even if you don't—it makes good sense to install a clock-type thermostat that will turn the heating unit on and off depending on the time of day. Most of these units require conversion to standard voltage. The economics of all this is discussed in Chapter 12, but for now take it as a "given" that such a thermostat will save precious fuel—and dollars.

In any case, simply replace a defective thermostat by unscrewing it from the wall and disconnecting the wires from the unit. Reconnect the new one to the same wiring, and you should have heat again (or "cool," if you have central air-conditioning).

Fig. 9-4. A transformer. All "hot" wires are inside the service panel. The wires you can see (and touch) are all low-voltage and safe to handle.

Transformers

Modern homes have a transformer in the basement mounted directly on the side of the entrance panel, or on a junction box not far away (Fig. 9-4). In homes without basements, transformers may be located in utility rooms, or nearby, somewhere close to the entrance panel. In older homes they may be more difficult to pinpoint, but look near the heating unit first, and follow the bell wiring to its source if you can't find them readily.

Makeup

Transformers have four terminals. Two are usually inside the junction box or entrance panel, and are connected to the source by standard 120-volt wires. The exterior (visible) terminals are connected to the low-voltage wiring systems. No connectors are needed for low-voltage wiring, and splices can be made anywhere.

A transformer is rarely at fault when there is a difficulty in the system. If a transformer fails, there will be a failure of both the chimes and the thermostat.

Replacement

To replace a transformer, determine the source of power. If it is connected to a junction box, deenergize the circuit to the box, as previously explained. Remove the transformer and replace with a new one having a secondary rating that matches the voltage required by the devices connected to it. Disconnect the 120-volt wires to the old transformer and reconnect to the new one. Do the same for the low-voltage wiring.

A transformer hooked up to the side of the entrance panel presents the same difficulties as any other connection to the service. While the chances of a serious injury are less than those for a new circuit, there is always the possibility of a stray hand touching hot wires.

Again, a professional electrician is in order for all but the bravest. If you feel that you must tackle the job yourself, note how the old transformer was connected, and follow a similar pattern for the new one. Connect the black wire to the fuse or breaker (usually in tandem with another black wire) and the white wire to the neutral bus bar.

Installing New Bell Wiring

There should rarely be any need to replace an entire low-voltage wire system. You may, however, wish to move the chimes or add a new set on another floor, thereby necessitating an exten-

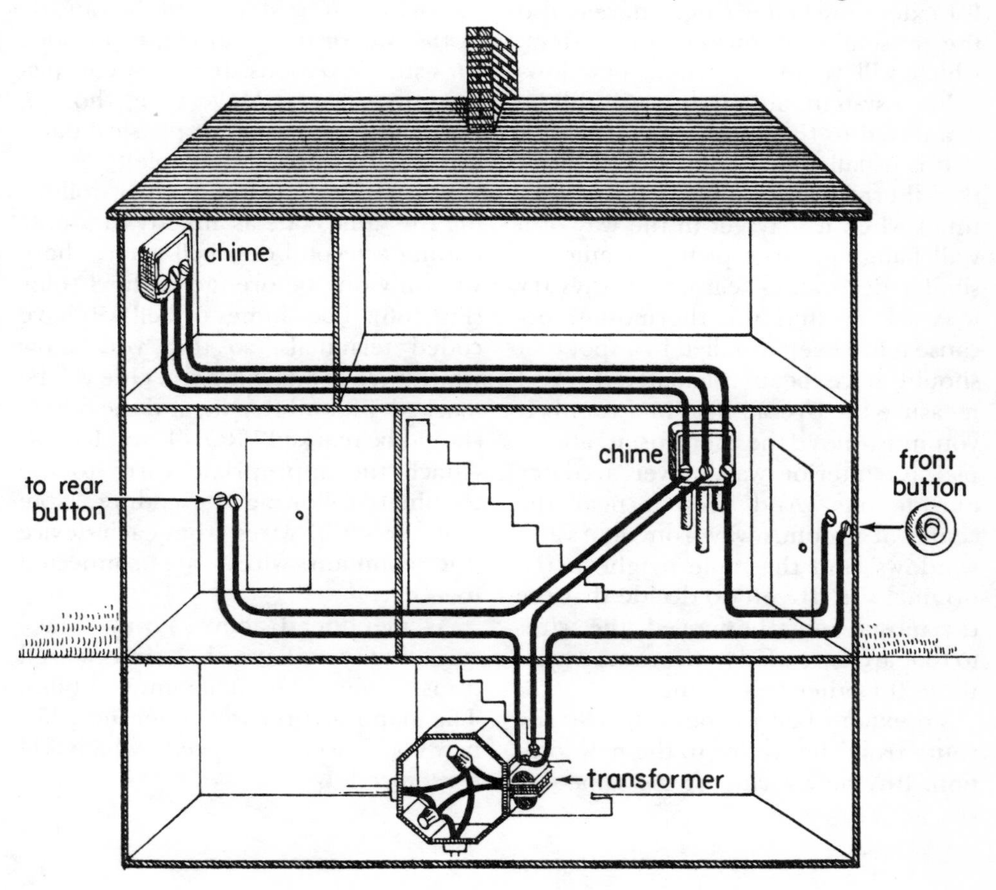

Fig. 9-5. Diagram for adding second set of chimes.

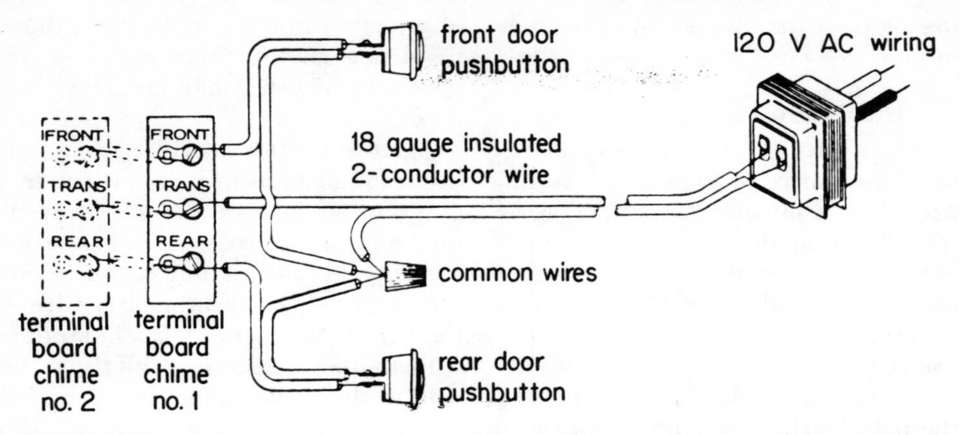

Fig. 9-6. Typical wiring set-up for front and rear chimes. Dotted lines show how to wire for additional chimes.

sion of the bell wires (Fig. 9-5). Adding an additional button to a door that does not have one is another occasion for extending bell wiring. There is also the possibility of an intercom system, which will require a whole new low-voltage system, although it will still be connected to the existing bell wire.

It is usually not a good idea to move the thermostat, although there are times when it may get in the way of a wall hanging, large picture frame, or similar decorating feature. Always try to work around the thermostat, because it has been installed in a spot that should have been calculated to best measure the overall heat in a room. If you must move the thermostat, always pick an interior wall, never a colder outside one. And keep it near the center of a room, away from doors and windows, and the same height as the original site. If you do decide that the thermostat must be moved, the wires to that are extended in the same way as those for other bell wiring.

To extend bell wiring, calculate the route from the source to the new location. Buy new wiring of the same size as the old, and connect the new to the old with a simple splice (acceptable for bell wiring). No box is needed. The new wire can be stapled to the outside of the wall or fished inside, if possible, for esthetic reasons. It is least conspicuous in corners. Make sure, though, that it is not exposed to physical damage (such as a careless staple).

Use different-colored wires, following the same code as already in use. If adding a second doorbell where there was only one before, get a third color (Fig. 9-6). The chimes or bell will have coded terminals, so that you know which wire to attach where. One will be marked TR or TRANS, and the others should be marked FRONT and REAR. Attach the appropriate wires to the terminals following the coding. Note that one set of wires from each device (the common wires) are connected together.

As mentioned above, always make sure that you have the correct-sized transformer when changing chimes. The manufacturer will either include a new transformer or specify which size is required.

Outdoor Lighting

Advantages of Outdoor Lighting

Safety

Outdoor lighting is an often neglected, but quite important, facet of home decoration. Indeed, it is more than just decorative. Proper outdoor lighting provides safety and security, as well as beauty and convenience. Dark recesses about the home are inviting to burglars and other assorted thugs. Dim stairways and walks are common sites for accidents to members of your family and others (and an occasion for lawsuits).

Decoration

In addition to basic needs for safety and security, however, outdoor wiring can provide a number of pluses. Lights can be used to highlight trees, shrubs, flower beds, and fountains. They can light pools, tennis courts, and other recreational areas. Patios and decks can be used 24 hours a day with proper lighting. Outlets can provide power for barbecue starters, electric lawnmowers, hedge clippers, and other outdoor equipment. Radios, television sets, and other amenities can transform your outdoor areas into true living space.

Special Equipment Needed

The modern approach to outdoor and landscaping lighting is to use low-voltage circuits and equipment which are not only safe, but are code-free on the secondary side.

Outdoor work is a completely new ball game as far as electrical equipment is concerned. Water is the traditional enemy of electrical equipment, and any exterior installation needs protection not only from wetness, but from corrosion, frost, animals, and other outdoor hazards.

Ground Fault Interrupter

Consequently, each electrical cable and device has to be specially designed to withstand the elements. Obviously, cable must be buried deeply enough to ward off the spades of landscapers and the claws of dogs in search of bones. Not so obvious, perhaps, is that it may have to be buried deeply enough to ward off the effects of frost—heaving of the earth, for example. The types of cable suited to underground use, and

Fig. 10-1. Outdoor lighting can make your home's entranceway safer and more attractive, and enable you to use outdoor recreation areas 24 hours a day. (*Courtesy of Western Wood Products Association*)

the depths at which they must be buried, are specified by the code. It is highly recommended that you check your local code for such information, as the requirements vary according to the specifics of the intended use of the installation. Remember, too, that since 1975 every exterior installation must be protected with a special safety switch—a ground-fault interrupter—that gives an extra measure of safety in case of a short or other circuit failure.

It was always an excellent idea to provide this additional safety device, and no one should try to short-circuit the Code in this respect. Technically a ground-fault *circuit* interrupter (GFCI), this device is usually referred to without the "circuit," and GFI is the more common abbreviation. It is what its name implies, a super-sensitive circuit breaker or fuse, that shuts down the circuit in about 1/40th of the time that a standard breaker would. Only 0.005 of an amp over the standard amperage trips the GFI, thus ensuring that a shock caused by a defect in the system will last only a tiny fraction of a second, and that serious injury or death can be prevented.

Cable

All fixtures and devices used outdoors must be specifically designed for such use (Fig. 10-2). Lights must have waterproof sockets, outlets must be protected by waterproof covers, and all boxes must be of weatherproof aluminum or steel (known as "T" or "PF" boxes). Switches and receptacles are the same as those used indoors, but waterproof covers and gaskets protect the wires inside from the elements. Cable must be specifically designed for outdoor use. Either nonmetallic type UF cable, which is encased in tough plastic, must be used, or the wires must

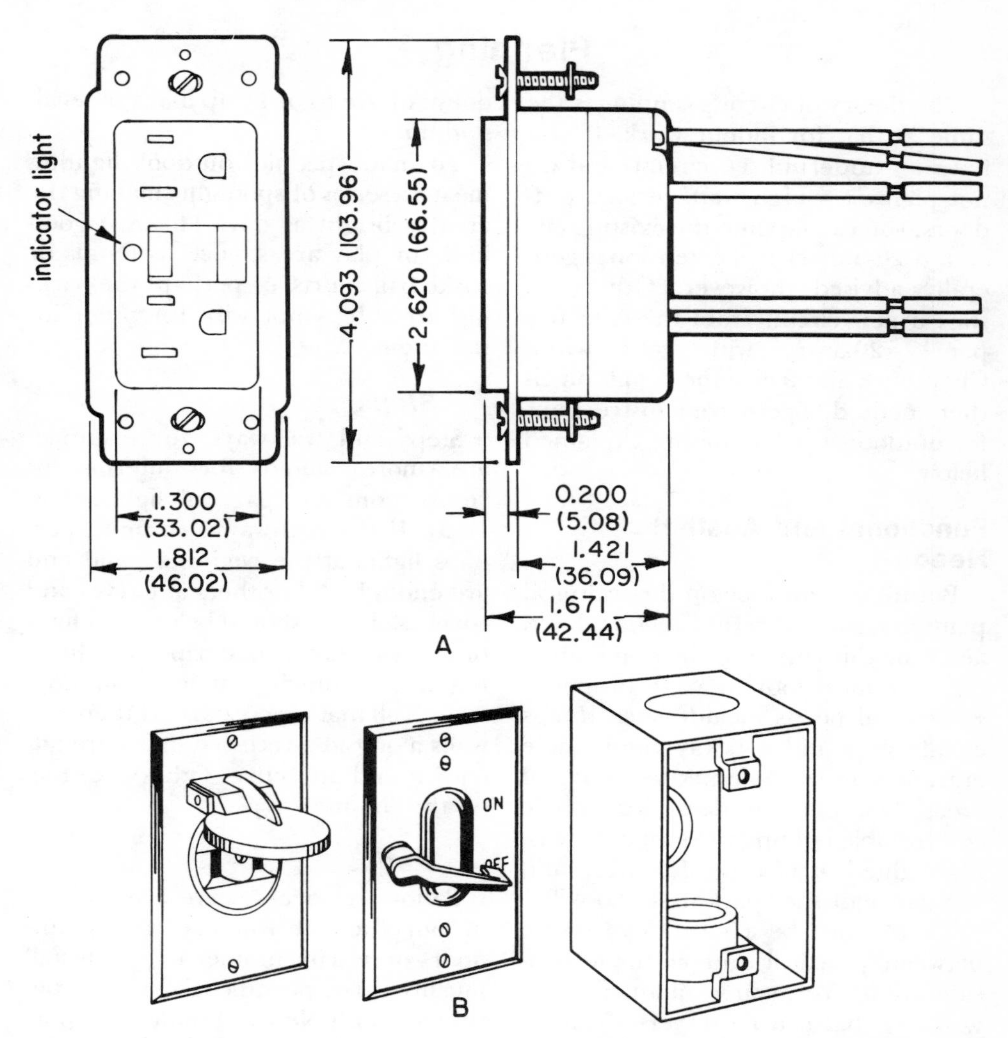

Fig. 10-2 (A) A GFI or ground-fault interrupter (measurements in parentheses are millimeters, all others are inches). (B) Specially designed exterior boxes and covers to keep wiring from the ravages of the elements.

be run through tough conduit (sometimes both, for convenience sake, as discussed below and shown in Figures 10-5 through 10-7). All aboveground wiring must be shielded in conduit, even if type UF cable is used.

Code Restrictions

Local codes can be even more restrictive than the NEC with regard to outdoor wiring, and the NEC is quite restrictive in itself. This is as it should be, because of the many hazards that outdoor installations can lead to.

Nevertheless, the basic techniques and overall circuit theory are the same as those for indoor work. If you understand the methods and principles discussed earlier in this book, you should be able to adapt them to outdoor work. The differences will be explained as we progress.

Planning

The theory of circuit planning is the same as that for indoor work. If you have an underutilized circuit, and do not plan on a high wattage load outdoors, you can tap into the existing circuit. A 20-amp circuit extension is generally advised, however. Otherwise, start a new circuit from the entrance panel (20-amp, with #12 wire). Chapters 8 and 9 give the details on either method. Specialized instructions for outdoor installations are explained below.

Functional and Aesthetic Needs

Before you even begin the technical planning, some careful functional and aesthetic thinking must be done. First, take a critical look at your property. Functional needs include such things as safe steps and walkways, and visible entrances and well-lighted recreational areas. Aesthetic considerations include comfortable lighting in lounging areas, highlighted landscape features, and avoidance of the "parking-lot look."

There must be a judicious balance between practical and aesthetic considerations. Yes, you want an entrance walkway that can be traversed safely, but you don't need to bathe the entire area in eye-straining glare. You want a patio where you can entertain, but it doesn't have to be lit up like a baseball stadium.

To many people, outdoor lighting means a series of spotlights making the area as bright as day. This *is* a good idea for play areas, such as tennis or basketball courts, or perhaps the parking area. It is not wise for other installations.

Steps

Steps and walkways, for example, are more suited for intermittent "mushroom" or "pagoda" lights, (Fig. 10-3). If the pathway is straight, perhaps lights at the beginning and end are enough. Add others at curves and corners. There should be enough light on steps so that no one trips over them, but not so much that it is blinding. That is all that is necessary. If there are walls alongside, recessed lights are adequate and attractive. Otherwise, rely on the "bonnet" type.

Patios

Patios and decks serve a multitude of purposes. If you like sitting outdoors on warm summer nights, install mushroom or pagoda lights on the perimeter, with plenty of outlets for portable lights, radio, TV, or barbecue spits. For parties, add spotlights. Even better, fit overhead lights with a dim-

Fig. 10-3. Outdoor lighting must be subdued and carefully designed. Here, the well-placed mushroom and spot lights illuminate flowers and shrubs while providing guidance for the flagstone walk. (Courtesy of General Electric.)

Fig. 10-4. This attractive garden can be enjoyed both day and night, from both inside the house and outdoors. (Courtesy of Westinghouse.)

mer control. That way you have full control over how much illumination you want. After all, with too much artificial light, you can't see the moon and the stars.

Trees and Shrubs

Bring out your creative best by highlighting trees, shrubs, and other landscape features. Use your imagination to conjure up special patterns, textures, and colors. One secret here is to place the lights so that they are themselves unobtrusive (Fig. 10-4).

Proper Placement

No matter where and how you place your outdoor lights, do all you can to keep them in the background. Avoid glare at all costs. If you light your entrance walk in such a way that the lights shine in your guests' faces, it is just as bad or worse than having no light at all.

Doing the Work

Once your plans are carefully drawn up, it's time to get to work. No matter how you slice it, outdoor work is expensive. Every device that is suited for the outdoors costs more than its interior counterpart. The need to keep everything weatherproofed necessarily raises the price. Even if you use low-voltage lights (pp. 120–121) the initial cost for the transformer is relatively high. (Additions to the system, however, cost less.)

Running Cable from Inside

The first step is to bring the current outdoors. You may be able to tap into an existing outdoor light, such as one located on the porch. In most cases, this will involve using an exterior box extender, because there will probably not be enough room in your present box for an additional set of wires. Since most of these lights are placed up high, however, you will have to run conduit all the way down to the ground, which is both unattractive and expensive.

The most common way of bringing the current outside is by either tapping into an underutilized circuit inside or by running a new circuit from the entrance panel. In either case, the cable must be somehow gotten through the exterior wall or foundation. (See p. 119 for running cable through masonry.)

Switches

The most desirable way to begin is by installing a switch inside the house near the outside door. If this is too difficult—and it often is—conduit may be run to the exterior, again near the door, and an outside switch installed. In any event, the GFI must be placed on the line somewhere before the first outside device.

GFI

Depending on the routing, an exterior or interior GFI is used. In the installation shown a new circuit was run from the entrance panel to the basement garage, where there was a receptacle-type, interior GFI. This provided a needed outlet in the garage. The cable was run directly from the GFI through the header joists directly above the sill plate. Use a ⅞-inch spade bit to drill through the header, and a small piece of ½-inch rigid conduit inserted in the hole from outside to the GFI. Type UF cable was inserted through the conduit to an elbow on the other end, where it ran through more conduit up to the exterior switch.

On the incoming "line" or "feed" side of the GFI, the white wire is attached to the white wire of standard type NM cable. The black wires are similarly attached. The corresponding wires on the outgoing "load" side of the GFI are attached to Type UF wiring.

When the wires to the GFI are attached properly, they are inserted into the box and covered with the special plate that comes with it. When the cable is attached to the entrance panel, the GFI becomes operational. If anything is wrong in the wiring on the line side, the GFI trips, shutting off the current on the line side of the circuit, as well as the integral outlet, if any. The GFI may be reset if there is only a temporary surge, by pressing the button so marked. If there is a short or other serious flaw, the GFI will keep tripping. There is also a test button to make sure the GFI is working properly.

Conduit

As mentioned above, all exposed outside work must be enclosed in conduit. Rigid metal or PVC plastic conduit is permitted by the NEC, but local codes should also be consulted. For the work described, the simplest method is to extend rigid metal conduit from the boxes into the ground, and use bare type UF cable underground for horizontal runs between the boxes.

Types of Cable

When deciding which type of underground cable to use yourself, you must balance several factors. The easiest material to work with is bare Type UF cable, but this must be buried below the frost line, and at least 2 feet below the surface (Fig. 10-5). If the ground is rocky, full of hard clay, or has a high water table, conduit may be a better choice. Rigid metal conduit need be buried only 6 inches below the surface, while PVC conduit should be buried a foot below the surface. (Thinwall EMT is not recommended for underground use.) Rigid metal costs quite a bit more than the other types, and is very difficult to bend. The PVC conduit bends more easily, but is more difficult to put together, requiring special connectors. (Rigid metal is prethreaded.) All types of conduit require that Type TW or UF wiring be pulled through after the conduit is laid.

It is impossible for anyone to dictate which method will work best for you. I live on very sandy Long Island ground, which is easy to dig. Even though I had to go down 3 feet to clear the frost line, I found it easier to do the digging than get involved in conduit.

(Instructions for working with conduit are given on pages 77–79. Type UF cable should be installed as flat and as straight as possible in the bottom of the trench as in Figure 10-5.) Where it was brought up to the boxes, I used lengths of rigid metal, with plastic bushings on the ends to avoid cutting the cable on the sharp end threads. (See Figures 10-6 through 10-11 for installation procedure.)

Exterior Boxes

Although many local codes forbid it, the NEC allows exterior "T" or "PF" boxes to be placed underground as long as they are not under sidewalks, driveways, and so on, and are easily identifiable and accessible. I therefore bought foot-long lengths of conduit to screw into the pagoda lamps and attached the boxes to the bottom of each. I bought boxes with two bottom "knockouts" (actually "screw-outs" on outside boxes). I used 2-foot lengths of ½-inch pipe in one hole for stability, and inserted the cable into the other knockout with special exterior connectors. These fittings have plastic bushings that are squeezed onto the cable inside the conduit fittings to make a weather-tight connection, while keeping the cable from slipping or getting damaged by the metal (Fig. 10-12).

A word of caution for those using exterior cable connectors. Always put

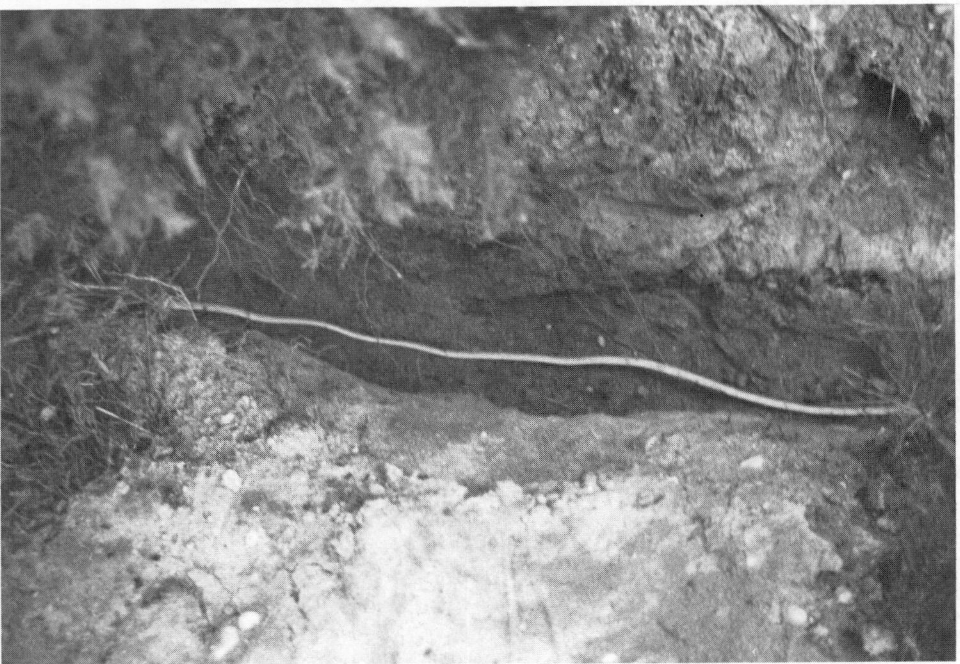

Fig. 10-5. (Above) Dig a trench deep enough to accommodate the Type UF cable without damage from the elements. It should be placed at least 2 feet below the surface. (Below) Stretch out the cable in the bottom of the trench, laying it as straight as possible. Backfill the trench after laying the cable to keep it in place.

Fig. 10-6. Strip Type UF cable like Type NM cable, but since it is considerably more difficult to free the wires from the solid plastic casing, use care and be patient.

Fig. 10-7. Weatherproof connectors consist of two pieces of metal with a plastic bushing inside.

Fig. 10-8. Tighten exterior connectors with a wrench.

Fig. 10-9. The incoming cable is already tightened at the back of the weatherproof box. At left is the outgoing cable with the connector separated to show the plastic bushing. Pipe at right is used to steady the box in the ground.

Fig. 10-10. A piece of 12-inch conduit runs from the box to the pagoda light. The fixture wires extend down through the conduit into the box.

Fig. 10-11. Completed fixture, with waterproof cover in place, can now be backfilled. Boxes can be placed underground only if they are accessible and readily identifiable.

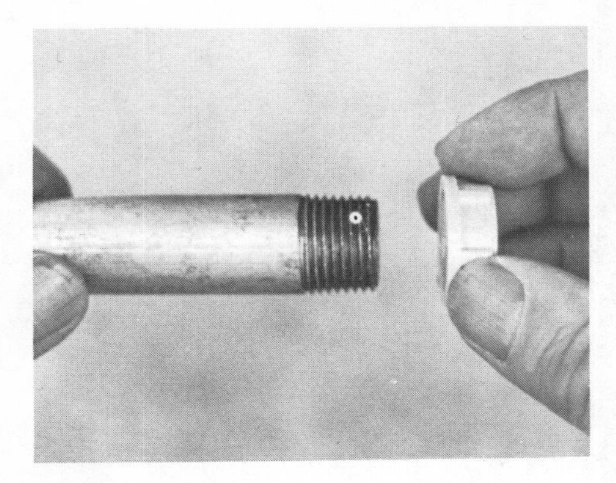

Fig. 10-12. If running UF cable directly into rigid metal conduit, use a plastic bushing over the threads of the conduit to avoid scraping and damage to the cable.

the cable inside the plastic bushing before attaching the other parts. Otherwise, you will have a very difficult time squeezing the cable through. Even then, it isn't easy assembling these fittings. I had to use a lubricant, and sometimes a hammer, to get all the parts together. They are supposed to be tight, and they are. Use a wrench to screw the assembled fittings to the box (Fig. 10-18).

Outside boxes are attached to walls, railroad ties, or whatever, by nailing or screwing through one of the punch-out screwholes in the back of the box. You will have to punch out the back of the box to get completely through. Rigid conduit actually holds the box in place, so don't worry about a little looseness when you attach the boxes. The conduit is secured by using conduit straps. You can actually install boxes without attaching them to anything, because the conduit will hold the box up. Where conduit does not extend far enough in the ground to give stability, drop a concrete block around the box and fill the hole with gravel. In Figure 10-16, you can see

Fig. 10-13. The receptacle box is screwed into the railroad tie. Rigid conduit is used whenever the cable is over ground, attached to the wood with straps. This shows the outlet attached, with the gasket in place over front of the box.

Fig. 10-14. Weatherproof cover goes over the gasket. Note tension-spring hinged covers over the outlets. These can be locked shut if advisable.

another example of improving outdoor lighting to accommodate both aesthetic and functional needs.

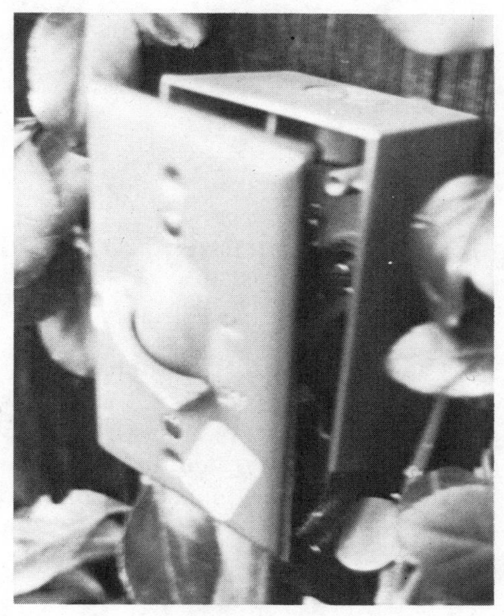

Fig. 10-15. A weatherproof switch cover just before it is screwed into the box.

Recessed Fixtures

To install a recessed fixture in a retaining wall, stairway, and so on, make a fixture-sized opening in the wall, using a cold chisel for concrete, brick, or other hard material. If the wall is thick, bore a ⅞-inch hole through to the back of the wall with a masonry bit or star drill. Install an outdoor box at the back of the wall, and connect the two with a nipple. Run the wires from the fixture to the box through the nipple and caulk any opening with mortar. Boxes can be recessed into concrete blocks or other masonry using similar techniques. Before drilling completely into a concrete block, make a test hole to make sure that you hit a hollow part of the block. Adjust the hole if necessary so that it fits into one of the cores instead of a solid section. Avoid the top row of a block foundation, because it is usually solid. The same applies for bringing cable through a foundation.

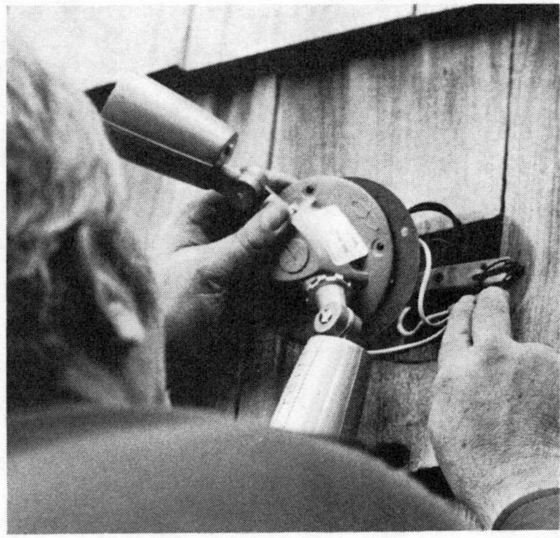

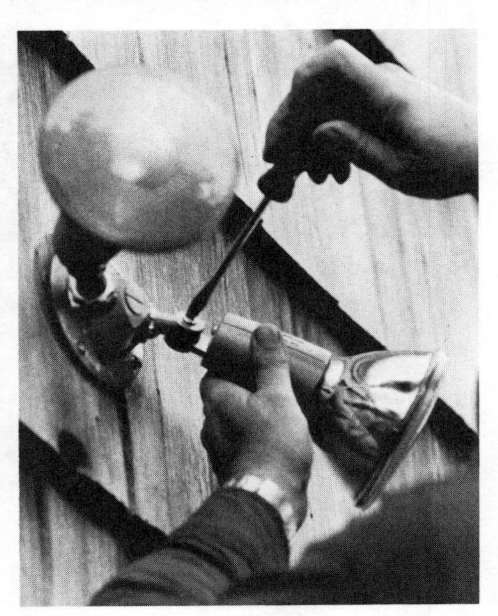

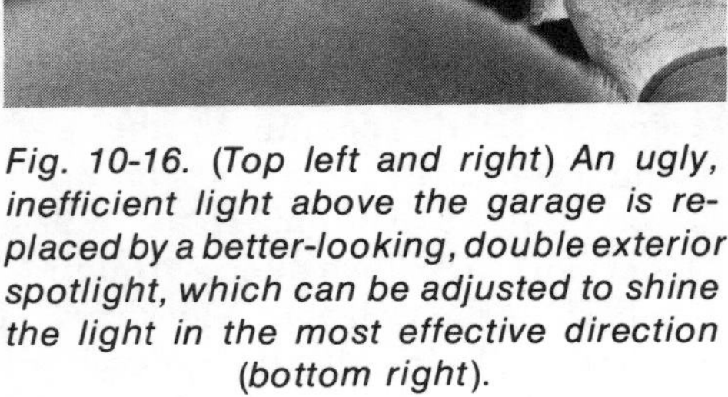

Fig. 10-16. (Top left and right) An ugly, inefficient light above the garage is replaced by a better-looking, double exterior spotlight, which can be adjusted to shine the light in the most effective direction (bottom right).

Using Low-Voltage Outdoor Lighting

Low-voltage outdoor lighting has most of the advantages of standard outdoor lighting and is quickly and easily installed with minimum risk. Like low-voltage indoor wiring, it operates from a transformer that steps down the voltage to a hazard-free 12 volts. The lighting is not as powerful as standard outdoor illumination, nor can receptacles for outdoor equipment be installed. For decorative or walkway lighting, however, it is an excellent choice.

Transformers

The expensive part of the installation is the transformer, which is usually sold as part of a kit with a half-dozen or so integrated lights. Some kits also contain a timer built into the transformer unit, so that the lights can be turned on and off automatically. Once you have the transformer, other lights can be added at minimal cost.

The transformer must be UL-listed for outdoor use. It is installed in the same way as any other outdoor fixture on the line side. Regular cable or conduit must be used up to the transformer, which is installed in the same way as an indoor transformer. On the load, or outgoing side, however, there are no restrictions as to where and how the wiring must be installed.

Technique

Follow manufacturer's instruction for running the wiring, installing lights, and so on. Most lights can be installed anywhere by spiking them into the ground. Wiring can usually be buried slightly underground, or laid on top of the ground, if desired. To avoid damage, however, it is best to run the wire alongside of a solid structure, such as a sidewalk, fence, or retaining wall. If wire is not included in the kit, check the instructions to make sure you use the correct size and type of cable.

Principles of Lighting

Proper home lighting is rarely seen as a major factor in decorating schemes, and is usually considered an expensive nuisance by builders. However, the judicious use of light should be a major ingredient in home design. Not only is it essential to work tasks, reading, dining, and almost everything else we do in the home, but intelligent lighting can help set a mood, highlight an important theme, and diminish decorating or building faults.

Think of the various rooms in your home. What are they used for? Should they be bright and airy, dark and mysterious, or somewhere in between? Perhaps there should be a little of each, as in the bedroom. When dressing, lots of illumination is essential. There are other times when darkness, or romantic dim light, better suits the occasion.

Importance of Adequate Light

Lighting is also vital in color and material considerations. Shiny metals, glass, and highly polished furniture or paneling can become uncomfortably glaring in strong light. Some colors can be lost in a dark room, or too bright in a well-lighted area. Dark walls, fabrics, and carpeting absorb a great deal of light, creating a need for more illumination.

Each room has its own particular problems, but let's take a typical foyer to illustrate the point. If it's like most other entranceways, there is one rather ugly "builder's fixture" in the center of the ceiling. For very small foyers, this may be adequate, or even too bright. If the room is a little larger, however, and contains a closet, or perhaps a painting or other wall decoration, you will undoubtedly be fumbling in the dark for your clothes and squinting to see what's on the wall. If the floor is marble, there will be a glaring bright spot in the center. Instead of, or perhaps in addition to, the center light, what you should have is a closet light and a spotlight on the wall decorations. Or, a simple change of bulb wattage may correct the problem.

In any case, if you look around your home with a critical eye toward the effects of the lighting, you will no doubt find numerous examples similar to the above. Lighting can, and should, be an integral part of home planning and design. Unfortunately, it seldom is. These deficiencies can be corrected, however, and sometimes amazing transformations can be performed if you understand and apply the principles of good lighting. But let's begin at the beginning.

Fig. 11-1. Objets d'art, a free-standing fireplace, and a very attractive ceiling fixture transform this large foyer into an attractive entrance. (*Courtesy of Condon-King*)

To apply the principles of good lighting in your home and to safeguard the eyesight of your family, you need to understand how you see. Seeing is not done by the eyes alone. The nerves, the muscles, and the brain are also involved.

Prolonged visual work in too little light or in glare can be as tiring as hard physical labor. The resulting exertion does not damage your eyes, but it can cause eyestrain and nervous fatigue. Glare is direct or reflected light that is visually disturbing or that interferes with seeing.

Consider the eye as a living camera that transforms light energy into sight. Light rays—from the sun, from a lamp, or reflected from any object—pass through the cornea and lens of the eye and are focused on the retina. The retina continually transfers pictures to the optic nerve and on to the brain. The final step in seeing is performed in the brain.

During your waking hours, this rapid, complex process constantly feeds back the shape, color, size, and motion of the things you see. More than three-fourths of all sensory impressions come through the eyes.

Four factors determine visibility (the ease and accuracy with which you see). They are:

Size: The larger the object, the easier it is to see.

Contrast: To be seen, an object must contrast to some extent with its background. A dark wood carving against a white wall can be easily seen, but a white figurine against a white wall is difficult to see.

Time: It takes time to see clearly. For this reason, rapidly moving blades of an electric fan blur. There is not enough time to see the individual blades.

Brightness: This is the amount of light the eye actually sees. It may be direct or reflected from objects and surfaces. There is no visibility in darkness and little in dim light. As light increases, surrounding surfaces become brighter and reflect more light to the eyes. As a result, you see more efficiently and effortlessly.

Of these factors, brightness is the only one over which you have much control. Ordinarily, you cannot change the size or form of an object, or slow down the speed with which it moves. In some cases, you can put an object against another background to give more contrast. But you *can* add more and better light to make seeing easier.

Where You Need Light

Does the lighting in your house make seeing easy wherever eyes do their work? In most homes, there are 30 or more seeing jobs that need specific lighting. (See checklist on p. 139).

To pinpoint the places in your home where light is needed, make a room-by-room check. Jot down the activities that take place in each room, remembering each member of the family as you do so.

The family living room, for example, is the setting for a wide range of activities. Here family members watch TV, entertain friends and relatives, read, study, telephone, write letters, or engage in other work or play that requires different kinds of lighting.

Once you are aware of the places where specific lighting is needed, you can plan your lighting for the activities that take place there. Good lighting, you will find, greatly increases the usable space within your home.

How Much Light Do You Need?

For comfortable, easy seeing throughout your home, you need to find out how much light is required in various areas. The output from fixtures, portable lamps, and built-in units in any room or area should provide general lighting, and also give the right kind and amount of lighting for specific activities. Natural window light should also be taken into consideration for daylight activities (Fig. 11-2.)

Lighting authorities of the Illuminating Engineering Society have determined the minimum amounts of light needed to do certain tasks. These

Fig. 11-2. This study area is a pleasure to use both day and night. Natural light provides plenty of illumination on bright days. The small lamp fills in at other times. (Courtesy of Nu-Tone)

levels of illumination are measured in footcandles, as shown in the table on page 140. A footcandle is the amount of light falling on a surface 1 foot away from the source of light of one standard candle (ordinary candle 1½ inch in diameter).

You can check your present lighting with a light meter. Compare your footcandle readings with the minimum levels recommended. Then increase the light if needed. (If you don't own a light meter, try to borrow one from a photographer. Perhaps the local lighting company or county extension can be of assistance.)

How To Get Good Quality Lighting

In an effectively lighted room, light is well distributed, free from glare, bright spots, and deep shadows. Large surfaces—walls, ceilings, and floors—should have favorable reflectances.

Light affects color. A lovely room by day can be even lovelier by night. Before you choose wall and fabric colors, try them under the same combination of lighting you will be using at home. (Some of the better clothing stores follow this principle.) Remember, too,

Fig. 11-3. The large window area here allows the use of dark colors in the rest of the room.

that dimming of incandescent bulbs changes the color of their light by making it more orange-red.

The light from some fluorescent tubes appears blue-white when combined with the yellowish light from incandescent lamp bulbs. To minimize color distortion, get deluxe fluorescent tubes in warm or cool white for both wall and ceiling lighting units. You may also buy fluorescent tubes that have the color spectrum of true sunlight—but at a price.

You'll find that light from warm white tubes blends well with light from incandescent bulbs and enhances complexions, foods, fabrics, and paints. Cool white tubes create a nice atmosphere and are effective in rooms with blue or green color schemes.

Favorable Light Reflectances

When planning the lighting of any interior, consider color and finish of walls, ceilings, wood floors or floor coverings, and large drapery areas. These large surfaces reflect and redistribute light within a room. Their lightness or darkness greatly affects the mood of a room.

White surfaces, of course, reflect the greatest amount of available light. Light tints of colors reflect light next best. Somber color tones absorb much of the light that falls upon them and reflect little light. (See reflectance table on p. 141).

If a large room gets plenty of daylight, you probably can use fairly strong color (Fig. 11-3). Light tints on walls make a small room seem larger.

Whatever the size of your room, try to keep colors within the 35 to 60 percent reflectance range. Ceilings should have reflectance values of 60 to 90 percent; floors at least 15 to 35 percent. Matte finishes (flat or low-gloss surfaces) on walls and ceilings diffuse light and reduce reflections of light sources. Glossy, highly polished or glazed surfaces produce reflected glare.

Light Sources

Incandescent bulbs and fluorescent tubes—in portable and wall lamps, in ceiling and wall fixtures, and in built-in lighting units—are the usual sources of electric light in homes (Fig. 11-4).

But bulbs and tubes do not ensure good lighting by themselves. You must select the right bulb or tube for the purpose you have in mind. Then the bulb or tube has to be placed in an appropriately designed lamp or fixture. And finally, the lamp or fixture must be correctly placed in the room.

Incandescent Bulbs

The incandescent bulb—the principal home light source for over 75 years—comes in a wide assortment of shapes, colors, sizes, and wattages.

General household bulbs, the most commonly used type, range from 15 to 300 watts. They are available in three finishes—inside frost, inside white (silica-coated), and clear. (See pp. 141–142 for selection guide for incandescent bulbs.)

Inside Frost and Inside White

The inside frost is the older bulb finish still in general use. Use bulbs of this type in well-shielded fixtures.

Bulbs with *inside white* finish (a milky-white coating) are preferred for many home uses. They produce diffused, soft light and help reduce bright spots in thin shielding materials.

Clear

Decoratively shaped *clear* bulbs add sparkle to chandeliers or dimmer-controlled simulated candles.

Three-way Bulbs

These bulbs have two filaments and require three-way sockets. Each filament can be operated separately or in combination. Make sure that a three-way bulb is tightened securely in the socket so that both contacts in the screw-in base are touching firmly.

The three lighting levels offered by these bulbs are particularly nice in portable lamps and pull-down fixtures.

Fig. 11-4. This handsome, paneled basement room mixes fluorescent lighting in the dining area with incandescent lights over the kitchen work area. (Courtesy Masonite)

You can turn the lamp high for reading and sewing, medium for viewing TV, conversation, or entertaining, and low for a night light or a soft, subdued atmosphere.

Dimmer switches are available for fixtures and many portable lamps. They make it possible to light from very low (nice for a romantic setting) to the maximum output of the bulb.

Tinted Bulbs

These bulbs create decorative effects indoors and outdoors. Silica coatings inside these bulbs produce delicate tints of colored light—pink, aqua, yellow, blue, and green. Home uses of these bulbs should be limited to lighting plantings, flowers, or art objects. You'll need to buy tinted bulbs of higher wattage because they give less light than white bulbs. Yellow bulbs used outside are less attractive to insects.

Silver-bowl Bulbs

These bulbs are standard household bulbs with a silver coating applied to the outside of the rounded end. They are used base up, and direct light upward onto the ceiling or into a reflector, using special ceiling fixtures. You can get them in 60-, 100-, 150-, and 200-watt sizes. They are generally found in basements, garages, or other work areas.

Reflector and Spotlight Bulbs

Reflector bulbs are available with silver coatings either on the inside or outside of the bulbs. *Spotlight bulbs* direct light

Fig. 11-5. This handsome, genuine oak bath has a well-lighted mirror, with fixtures on both sides using insidefrost bulbs. (Courtesy of Nu-Tone)

in a narrow beam and generally accent objects.

Heat-resistant Bulbs

These bulbs are used outdoors. They are not affected by rain and snow. Common sizes are 75 watts, 150 watts, and up.

Decorative Bulbs

These bulbs are designed to replace bare bulbs in older fixtures and sockets (Fig. 11-6). Some shapes and sizes are made for traditional fixtures (chandeliers and wall sconces); others combine contemporary styling and function. Bulb shapes include globe, flame, cone, mushroom, and tubular.

Some of these bulbs are made of diffusing type glass and are tinted to produce colored lighting effects. Clear bulbs are used to produce sparkle in crystal chandeliers. When selected to harmonize with fixtures and room decor, decorative bulbs offer a pleasing, low-cost solution to a lighting problem.

Floodlight Bulbs

Floodlight bulbs spread light over a larger area, and are suitable for floodlighting horizontal or vertical surfaces. Typical floodlight sizes include 30, 50, 75, and 150 watts.

Colored floodlight bulbs are available for indoor or outdoor use. The tints—particularly pink and blue-white—create nice effects on planters or flowers and are acceptable for lighting people and furnishings. Strong colors—blue, green, and red—are best reserved for holiday and party decorations.

Fluorescent Tubes

Most households have a need for fluorescent lighting in some form. (For guidance in selecting fluorescent tubes and fixtures, see the selection guide on p. 142.)

The whiteness of a standard tube is indicated by letters, WW for warm white; CW for cool white. (The addition of an "X" to these letters indicates a deluxe tube. Deluxe tubes are designed to bring out certain colors with

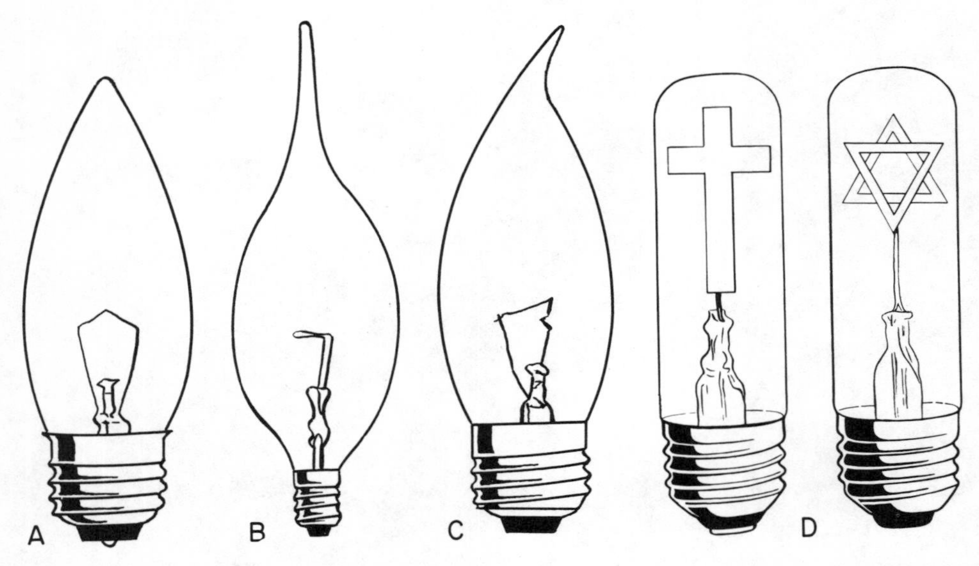

Fig. 11-6. Decorative bulbs come in many sizes, in either standard or candelabra bases.

maximum intensity, but may produce 30 percent less light per watt than the other types.)

A warm white tube gives a flattering light, can be used with incandescent light, and does not distort colors any more than incandescent light does. A cool white tube simulates daylight (but not very well) and goes nicely with cool color schemes of blue and green. Deluxe tubes are recommended for home use, but may be difficult to find.

Types of Lamps

Portable Lamps

Well-designed floor and table lamps may be difficult to find, but are worth looking for. Try to find lamps that combine function and beauty, and pay close attention to what's under the shade. Choose lamps that make seeing comfortable and, at the same time, harmonize with your furnishings, color schemes, and with other lamps and accessories.

Design

Generally speaking, the design of a lamp should be akin to the style and decoration of the room, its scale in accord with the furniture it appears on or over, the amount of light it radiates must suit the purpose intended.

Dimmer-switch controlled lamps can be purchased, or converted from standard lamps by adding a rheostat control. These give greater flexibility than three-way lamps that use three-way sockets and require three-way bulbs.

Small high-intensity lamps are not designed for study, reading, or general work. They can, however, provide a concentrated area of high-level light for special tasks, such as sewing, crafts, or fine-detail work. They should always be used in combination with good general lighting.

Uses

If you select a lamp for style or color alone, do not expect to use it for close work. It is a decorative lamp. Use it to:

- Give limited general lightning.
- Brighten a corner, foyer, or hallway.
- Display an object of art or an accessory.

Any portable lamp, regardless of size, should be sturdily built. See that the power cord is well-protected from sharp edges where it enters the lamp base and the vertical pipe.

Table Lamps

Correct Placement

Before you shop for table lamps, jot down the heights of tables on which lamps will be placed, and the height of any chair or sofa seat on which a person using the lamp will be seated. Take these figures with you as you shop. You may also want to consider the eye level height of persons using a lamp, particularly if an individual is unusually tall or short. Table height plus lamp base height (to the lower edge of the shade) should equal the eye height of the person using the lamp.

Eye height, of course, depends on the height of the chair or sofa seat and the eye level of the person seated. Eye height is usually 40 to 42 inches above the floor. For comfortable seeing, the bottom of the lampshade should also be 40 to 42 inches above the floor (Fig. 11-8).

Other points to consider in selecting and using table lamps are:

Shade dimensions: Shades on lamps for reading, sewing, or studying should be 16 inches wide or more at the bottom, 9 inches wide at the top, and at least 10 inches deep.

Minimum bulb wattages: Single-socket

Fig. 11-7. Lamps should be chosen to fit in with their surroundings. These modern Schiller-Cordey chrome lamps complement the contemporary setting perfectly.

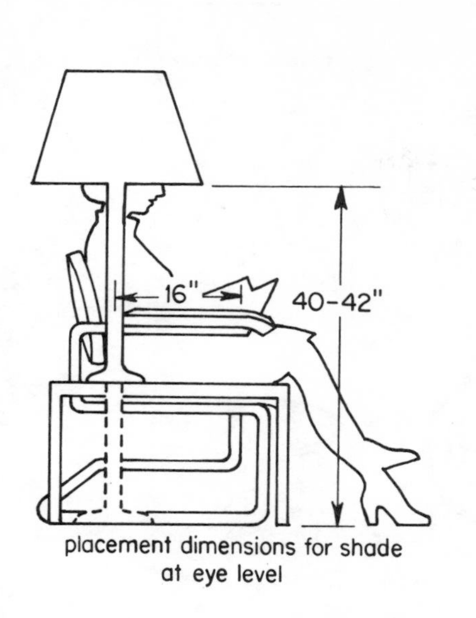

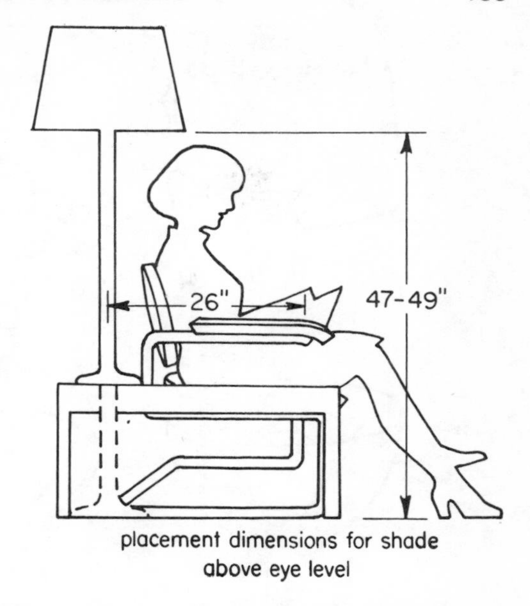

Fig. 11-8. No matter what type or height of the lamp, the bottom of the shade is 40–42 inches above the floor for close work or 47–49 inches when the light is further away.

lamps for reading, sewing, or studying should be 150 watts. Multiple-socket lamps should have at least three 60-watt bulbs.

Sockets: The center of the light source (bulb) should be located in the lower third of the shade.

Diffusers: Diffusers soften and spread light. Some types are illustrated in Figure 11-9.

Floor Lamps

In choosing a floor lamp, keep in mind exactly where it is to be located in your home. Floor lamps should harmonize with furnishings, and be carefully scaled to space. Choose lamps sized and constructed for proper placement without interfering with house traffic.

Small floor lamps—standard, swing arm, or bridge-type—may be 43 to 47 inches from the floor to the bottom of the shade. Large lamps measure 47 to 49 inches from the floor to the bottom of the shade.

For reading, a floor lamp with a fixed or swing arm is correctly placed when the light comes from behind the shoulder of the reader, near the rear of the chair—either at the right or the left—but never from directly behind the chair.

If a floor lamp is used for prolonged reading or sewing, it should have a bulb wattage of 200 or 300 watts; the minimum bulb wattage for reading is 150 watts.

Desk and Study Lamps

A well-lighted desk or study center makes it easy to concentrate. Good-quality light falls on the task at hand. Eyestrain and physical tension are reduced to a minimum because the worker has enough light and little or no glare.

For continued study or deskwork, be sure your lighting arrangement gives an average of 70 footcandles over the work area. A lamp correctly placed and fitted with a 200-watt bulb will give

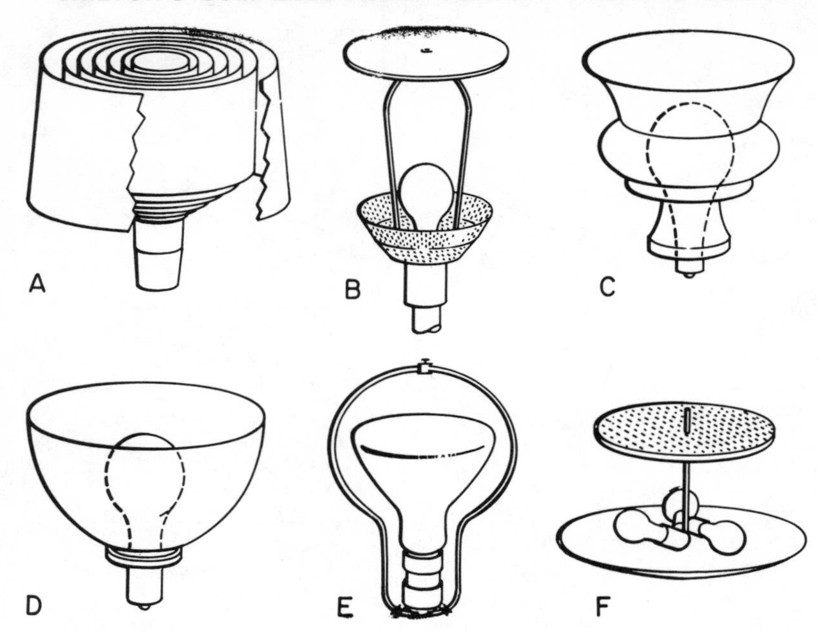

Fig. 11-9. Diffusers come in various styles and shapes. (A) a molded prismatic diffuser with a louvered top shield (B) a prismatic refractor at the base of the bulb, (C) a glass top diffuser, (D) a white glass bowl diffuser, (E) a self-diffusing R-40 white indirect bulb, and (F) a plastic diffusing disc on the bottom with a perforated disc at the top.

this amount of light. If you do not have a diffuser on the lamp, using a white bulb helps to reduce glare.

Several lamp manufacturers make specially designed, higher wattage study lamps. These improved lamps reduce glare to a minimum. They also give nearly twice as much light as other study lamps and spread light over a wider area. For these lamps, 200-watt bulbs are recommended.

Placement

A study lamp should be placed about 12 inches back from the front of a desk or table, about 15 inches from the work. This gives excellent lighting for intensive and prolonged deskwork.

A swing arm on a floor lamp makes it possible to place light in good position for study or deskwork. Such a floor lamp should measure 47 to 49 inches from the floor to the bottom of the shade. It is usually fitted with a white glass diffusion bowl and a large socket for a 100/200/300-watt three-way bulb, a medium socket for a 50/100/150 or 50/200/250 three-way bulb, or a single 150-watt bulb.

If you plan a study area under a wall bracket or pin-up-type lamp, choose a bracket with a swing arm so you can move the center of the shade forward. Then you can position the light to your best advantage.

Here are other suggestions to improve light in a study center:

- Choose a desk with a light-colored, nonglossy finish.
- Use a light-colored blotter or desk pad.
- Paint walls in neutral or light colors or select a plain wallpaper or one with a small, quiet design.

Shades and Bases

Shades

Shades made of translucent materials are usually chosen for portable lamps. Desk lamps use slightly translucent shades to avoid uncomfortable brightness in the eyes of the person using the lamp. Shades for dressing table or dresser lamps should be highly translucent because light reaching the face must come through the shade. Shade materials suitable for vanity lamps are too translucent for reading, study, and decorative lamps.

For effective light reflectance, the inside of the shade should be white or near-white. Good color choices for the outside of the lampshade are neutral or pale tints, off-white, beige, and light gray. Try to avoid excessive contrast between the color of the shade and adjacent walls.

Shape

The shape of a lampshade also affects lighting. Straight or nearly straight lines are preferable to extreme curves. Look for a shade that gives a wide spread of downward light as well as some upward light.

Base

The base of a portable lamp needs to be heavy enough to support the lamp firmly and keep it from upsetting easily. The design of the base should be appropriate to its function. It should please, but not distract. Grotesquely shaped bases are seldom in good taste.

Materials in lamp bases often relate to certain furniture styles or periods. Some popular furniture styles and appropriate base materials include: early American—pewter, brass, stoneware, copper, wrought iron, pottery, wood; 18th century or traditional—silver, porcelain, china, cloisonné, crystal, marble; contemporary—metals, glass, wood, cork, ceramics.

Diffusers and Shields

Diffusers are bowl- or disc-shaped devices that surround the lamp bulb under the shade. They scatter and redirect light, soften shadows, and reduce reflected glare.

Effective diffusion materials in order of preference include blown milky glass, enameled glass, flashed opal, and plastics. Undershade diffusers are available for use in study and reading lamps. One is a highly reflective, inverted metal cone. Other diffusers are bowl-shaped, prismatic reflectors. Shields are also used to prevent glare.

Perforated metal shields or plastic louvered shields, placed above the bulb, keep direct glare from reaching the eyes of the passerby. The mushroom-shaped (R-40 type) 150-watt bulb made of white diffusing glass needs no diffuser or shield. It serves well for casual reading. See Figure 11-9.

How to Improve Present Lighting

When you redecorate, finish walls in light pastel colors, and ceilings in white or near white or a pale tint. Flat or low-gloss paint on walls and ceilings helps diffuse light and makes lighting more comfortable. Use sheer curtains or draperies in light or pastel tints.

Improving Balance

Add portable lamps for better balance of room lighting. Install structural lighting (valance and cornice) in living areas where there is only one ceiling light or none. Eight to 20 feet of wall lighting will add a feeling of

Fig. 11-10. Other places that need a lot of special lights are the bridge table shown, and the kitchen-bar, which has a series of small spots. Note the modern multilight table lamp in the sitting area. (*Courtesy of Armstrong*)

spaciousness to an average-sized room and make the lighting more flexible.

Replace present bulbs with those of higher wattage, but do not exceed the rated wattage of the fixture. A minimum of 150 watts is needed in many single-socket lamps. For better control of lighting, use three-way bulbs or dimmer switches. Standard, one-way sockets can be changed to accommodate three-way bulbs. If you want a higher-wattage fluorescent unit, the fixture must be changed (see page 46). The light output of a fluorescent unit is about three times that of an incandescent unit of the same wattage.

Improving Efficiency

For efficiency, use one large bulb rather than several small ones. A 100-watt bulb gives as much light as six 25-watt bulbs, but only uses about two-thirds as much current.

Replace outmoded bare-bulb fixtures with well-shielded ones. Cover all bare bulbs or tubes in a ceiling fixture with a shade or diffuser. Some of these diffusers clip to the bulb. Others hang from small chains attached to the body of the fixture. Large diffusers, sometimes called adapters, may have supporting frames that are screwed on the sockets of single-bulb fixtures. An inexpensive way to avoid the glare of bare bulbs in a ceiling fixture is to replace these bulbs with silver bowl bulbs or decorative mushroom-shaped R-40 bulbs.

Increasing the Number of Outlets

Put in additional convenience outlets if needed for correct lamp placement. Added outlets prevent overloading through the use of dangerous octopi plugs and extension cords. Surface wiring strips may be attached along baseboards or counter tops, if allowed by local codes. These strips may be more economical than adding built-in convenience outlets. Be sure any surface wiring system you choose is of the correct size and carries the Underwriters Laboratories seal of approval.

Replacing shades

Lift the lampshade on a portable lamp with a riser if the bulb is too high. When a bulb is too high it restricts the downward circle of light and shines into the eyes of persons standing near. Risers come in multiples of ½ inch and can be screwed to the top of the harp to lift the shade the amount needed.

Replace the lampshade with a deeper shade if bulb is too low in lamp and bulb shows beneath lower edge of shade. Or, if you prefer, use a shorter harp or a different diffusing bowl.

If a lamp base is too short, set the base on wood, marble, ceramic, or metal blocks to raise lamp to proper height. For ease in handling, cement the block to the base.

Get replacement shades for table lamps if present shades do not meet specifications. Choose shades made of translucent materials with white linings and open tops. (For shade selection, see page 135).

Rewiring Sockets

Invert and rewire the socket of old-style bridge lamps. Then add 6-inch diffusing bowls and larger, wider shades that give softer, better lighting.

A Guide to Lighting Fixtures

Lighting fixtures usually provide the general lighting in a home. When they are well-chosen, they also add decorative tone and a pleasant atmosphere.

The basic principles of lighting—quantity, quality, color, and reflectance of light—should be considered in selecting fixtures. Individual fixtures may be combined with structural lighting for pleasing effects. And don't neglect spots and other specialized lighting (Fig. 11-10).

The manufacturers' wattage rating and the size of the fixture must be large enough to accommodate the largest wattage bulb needed to light the area. Often, you need more than one fixture. For example, a large rectangular kitchen may need two 48-inch two-tube fixtures placed end to end.

Check fixtures carefully before buying them. Here are some points to keep in mind:

- Incandescent bulbs should be no closer than ¼ inch to enclosing globes or diffusing shields.
- Top or side ventilation is desirable in a fixture to keep temperatures low and to extend bulb life.
- Inside surfaces of shades should be of polished material or finished with white enamel.
- Shape and dimension of a fixture should help direct light efficiently and uniformly over the area to be lighted.
- Plain or textured glass or plastic is

preferred for enclosures and shades.

- The general guidelines for portable lamps and shades also apply to fixtures.

Dimmers

Dimmer controls add convenience, safety, and flexibility to your home lighting in bedrooms, bathrooms, halls, and living rooms. Gradations of light—from full bright to very dim—are possible simply by turning a knob.

A low level of lighting is helpful in the care of small children, sick persons, and others who need assistance during the night. House guests, who are unfamiliar with their surroundings, appreciate night lights.

You can make dramatic changes in the mood of a room by softening lights with a dimmer switch. Lights can be lowered when listening to music, enjoying a fire on the hearth, or having a quiet dinner.

Dimmer controls for incandescent bulbs are simple, compact, and can be mounted in walls in much the same way as off-on switches. Dimmer switches are now available, too, for portable lamps. Be sure that the wattage capacity of the dimmer control is equal to or more than the total wattage to be controlled.

Dimmer-controlled fluorescent fixtures must be preplanned before installation. They are considerably more expensive than incandescent dimmer units. The control combines with a special built-in ballast, and can operate one or more specially designed fluorescent fixtures as a unit.

Track Lighting

Advances in track lighting design over the past years have made this type of lighting more attractive for home use today than it was formerly. New innovations such as the octopus outlet make it easy to attach the tracks to a ceiling junction box. From the octopus, tracks can be extended in any direction.

Track lights are used to excellent effect to light up dark room corners, provide light for plants, illuminate hallways or serve as dramatic accents to favorite visual effects such as painting, fireplace, or a piece of sculpture.

The individual lights can be secured in the track anywhere, and come in various styles and shapes (as do the tracks). Various wattages can be bought, or the lights can be dimmer-controlled to provide the degree of illumination desired.

Track lighting is made by several different manufacturers, and each design is different. See your electrical supply dealer for information on how to choose and install the various systems.

Lighting Maintenance

Home lighting equipment needs regular care and cleaning to keep it operating efficiently. A collection of dirt and dust on bulbs, tubes, diffusion bowls, lampshades, and fixtures can cause a substantial loss in light output.

It's a good idea to clean all lighting equipment at least four to six times a year—bowl-type portable lamps should be cleaned monthly.

Here are some suggestions for taking care of lamps and electrical parts:

- Wash glass and plastic diffusers and shields in a detergent solution, rinse in clear warm water, and dry.
- Wipe bulbs and tubes with a damp, soapy cloth, and dry well.
- Dust wood and metal lamp bases with a soft cloth and apply a thin coat of wax. Glass, pottery, marble, chrome, and onyx bases can be washed with a damp soapy cloth, dried, and waxed.
- Clean lampshades with a vacuum cleaner using a soft brush attachment, or dry-clean them. Wash silk or rayon shades that are hand sewn to the frame, with no glued trimmings, in mild, lukewarm suds, and rinse in clear water. Dry shades quickly to prevent rusting of frames.
- Wipe parchment shades with a dry cloth.
- Remove plastic wrappings from lampshades before using. Wrappings create glare and may warp the frame and wrinkle the shade fabric. Some are fire hazards.
- Replace all darkened bulbs. A darkened bulb can reduce light output 25 to 50 percent, but uses almost the same amount of current as a new bulb operating at correct wattage. Darkened bulbs may be used in closets or hallways where less light is needed.
- Replace fluorescent tubes that flicker and any tubes that have darkened ends.

Checklist of Activities

Living areas (living room, dining room, family or recreation room):
Reading or studying (prolonged).
Reading (casual, intermittent).
Viewing television.
Visiting and conversation.
Playing games (adult, children).
Reading music.
Setting table.
Dining.

Work areas (kitchen, laundry, workroom, home office):
Reading recipes and measuring ingredients.
Reading labels and following directions.
Inspecting and sorting foods.
Reading dials and checking foods as they cook.
Washing dishes and cleaning equipment.
Sorting and pretreating laundry.
Ironing or pressing.
Sewing (reading directions, cutting, fitting, machine and hand stitching).
Repairing small appliances.
Working on hobbies (artwork, collections, model building, photography).
Housekeeping (dusting, vacuuming, waxing, washing walls, woodwork, and other surfaces).
Deskwork (record-keeping, studying, typing).

Private areas (bedrooms, bathrooms):
Personal grooming (bathing, shaving, manicuring, shampooing and arranging hair, applying cosmetics).
Assembling clothes from closet or storage unit.
Dressing.
Caring for infants, small children, or the sick.
Reading medicine labels or taking temperatures.
Reading in bed or at bedside.

Halls, stairways, entrances:
Moving between rooms.
Using steps and stairs.
Reading house numbers.
Identifying callers.
Safe access to house or garage.

Minimum Levels of Illumination

(Recommended by Illuminating Engineering Society)

Specific Visual Task	*Amount of Light on Task in Footcandles*
Reading and writing:	
Handwriting, indistinct print, or poor copies	70
Books, magazines, newspapers	30
Music scores, advanced	70
Music scores, simple	30
Studying at desk	70
Recreation:	
Playing cards, table games, billiards	30
Table tennis	20
Grooming:	
Shaving, combing hair, applying makeup	50
Kitchen work:	
At sink	70
At range	50
At work counters	50
Laundering jobs:	
At washer	50
At ironing board	50
At ironer	50
Sewing:	
Dark fabrics (fine detail, low contrast)	200
Prolonged periods (light-to-medium fabrics)	100
Occasional (light-colored fabrics)	50
Occasional (coarse thread, large stitches, high contrast of thread to fabric)	30
Handicraft:	
Close work (reading diagrams and blueprints, fine finishing)	100
Cabinetmaking, planing, sanding, glueing	50
Measuring, sawing, assembling, repairing	50

General Lighting	*Average Light Throughout Area in Footcandles*
Any area involving a visual task	30
For safety in passage areas	10
Areas used mostly for relaxation, recreation, and conversation	10

Reflectance Values of Common Finishes

Color	Approximate Percent Reflection
Whites:	
Dull or flat white	75–90
Light tints:	
Cream or eggshell	79
Ivory	75
Pale pink and pale yellow	75–80
Light green, light blue, light orchid	70–75
Soft pink and light peach	69
Light beige or pale gray	70
Medium tones:	
Apricot	56–62
Pink	64
Tan, yellow-gold	55
Light gray	35–50
Medium turquoise	40
Medium light blue	42
Yellow-green	45
Old gold and pumpkin	34
Rose	29
Deep tones:	
Cocoa brown and mauve	24
Medium green and medium blue	21
Medium gray	20
Unsuitably dark colors:	
Dark brown and dark gray	10–15
Olive green	12
Dark blue, blue-green	5–10
Forest green	7
Natural wood tones:	
Birch and beech	35–50
Light maple	25–35
Light oak	25–35
Dark oak and cherry	10–15
Black walnut and mahogany	5–15

Selection Guide for Incandescent Bulbs

Activity	Minimum Wattage (inside white)
Reading, writing, sewing:	
Occasional periods	150
Prolonged periods	200 or 300

Selection Guide for Incandescent Bulbs (continued)

Activity	*Minimum Wattage (inside white)*
Grooming:	
Bathroom mirror:	
One fixture each side of mirror	one 75 or two 40
One cup-type fixture over mirror	100
One fixture over mirror	150
Bathroom ceiling fixture	150
Vanity table lamps, in pairs (person seated)	100 each
Dresser lamps, in pairs (person standing)	150 each
Kitchen work:	
Ceiling fixture (two or more in a large area)	150 or 200
Fixture over sink	150
Fixture for eating area (separate from workspace)	150
Shopwork:	
Fixture for workbench	150

Selection Guide for Fluorescent Tubes

T12 (1½-inch diameter) Tubes

Use	*Wattage and Color (WW = warm white, CW = cool white)*
Reading, writing, sewing:	
Occasional	One 40w or two 20w, WW or CW
Prolonged	Two 40w or two 30w, WW or CW
Wall lighting (valances, brackets, cornices):	
Small living area (8-foot minimum)	Two 40w, WW or CW
Large living area (16-foot minimum)	Four 40w, WW or CW
Grooming:	
Bathroom mirror:	
One fixture each side of mirror	Two 20w or two 30w, WW
One fixture over mirror	One 40w, WW or CW
Bathroom ceiling fixture	One 40w, WW
Luminous ceiling	For 2-foot squares, four 20w, WW or CW 3-foot squares, four 30w, WW or CW 4-foot squares, four 40w, WW or CW 6-foot squares, six to eight 40w, WW or CW
Kitchen work:	
Ceiling fixture	Two 40w or two 30w, WW
Over sink	Two 40w or two 30w, WW or CW
Counter top lighting	20w or 40w to fill length, WW
Dining area (separate from kitchen)	15 or 20 watts for each 30 inches of longest dimension of room area, WW
Home workshop	Two 40w, CW or WW

Teleview in the Right Light

Viewing television in a darkened room is extremely tiring to the eyes because of the sharp contrast between the bright screen and unlighted surroundings. To avoid eyestrain and fatigue, provide a low to moderate level of lighting throughout the viewing area.

Wall lighting from valances and brackets creates a delightful background for watching television. When you use these types of lighting, position your TV set to the side or in front of the lighted walls.

Another way to offset the brightness of the screen and make viewing comfortable is to place one or two portable lamps behind or at the sides of the set. This helps prevent reflections on the TV screen. If the lamps have three-way controls, turn them on the low settings.

Pros and Cons of Light Sources

Incandescent Bulbs	*Fluorescent Tubes*
Can be concentrated over a limited area or spread over a wide area.	Provide more diffused lighting—a line of light, not a spot.
Initial cost less than fluorescent tubes.	Higher initial cost, but greater light efficiency—three to four times as much light per watt of electricity.
Designed to operate at high temperature.	Cool operating temperature. Generally about one-fifth as hot as incandescent bulbs.
Have average life of 750 to 1,000 hours.	Operate seven to ten times longer than incandescent bulbs.
Wattages range from 15 to 300 watts.	Wattages for home use range from 14 watts (15 inches long) to 40 watts (48 inches long).
Amount of light can be increased or decreased by changing to bulbs of different wattage because most bulbs have same size base.	Cannot be replaced by higher or lower wattage tubes.
Require no ballast or starter.	Require ballasts, and in some cases, starters.
Do not interfere with radio reception.	May cause noise interference with radio reception within 10 feet of the tube location.
Suitable for use in less expensive fixtures.	Adaptable to and commonly used in custom-designed installations and in surface-mounted and recessed fixtures.
Available in colors to enliven decor and accessories. Colored bulbs are 25 to 50 percent less efficient than white bulbs.	Available in many colors (plus deluxe cool white and deluxe warm white) at much higher light output than colored incandescents.
Gain flexibility by use of three-way bulbs and multiple-switch controls or dimmer controls designed for incandescent bulbs.	Gain flexibility by use of dimming ballasts combined with dimmer controls designed for fluorescent tubes.

Electricity, Energy, and Money

The Energy Crunch

Yes, Virginia, the energy crunch is real. The United States and many other non-OPEC countries are facing a bleak economic future because of the scarcity of oil.

Many of us simply refuse to accept this fact. It's understandable, too, because we have been deceived by both government and big business for so long that we have difficulty believing anything we're told. Even if we don't take their words at face value, however, pure logic should tell us that we cannot continue to greedily gulp up oil (or any other natural resource) forever. Oil is a nonrenewable resource, and there is just so much of it in the ground. Even if we find more oil fields, or economically feasible ways to convert oil shale, sand tars, and coal into oil, it will take a long time, and these sources will some day run out, too.

What is important is that our dependence on imported oil is somewhat akin to other countries' dependence on imported wheat or military hardware. It costs dearly.

Even more important, for the present at least, is that our dependence on foreign oil contributes heavily to inflation, a shrinking dollar, and consequent economic stagnation. Our entire way of life is threatened. We have already learned that we can no longer just jump in our cars and go wherever we want without thinking of gasoline, its price and availability. The costs of owning and even renting a home are soaring because of dramatically rising heating-oil costs. Again, another symbol of prosperity is threatened.

Is there anything we can do about it? Yes, there is. We can invest in crash-type programs to find alternative sources of energy. But that isn't all. As a nation, we import 40 percent of the world's exportable oil. That's greedy, as we have been reminded by other importing nations. It's almost obscene. We're hooked on oil, and we have to stop, cold turkey if need be.

We *must* conserve. There are many ways that this can be done, but since this book is about electricity, I'll concentrate on that. Actually, there's more that can be done in other ways to conserve oil, such as simply turning down thermostats and wearing sweaters—or using more wood and coal. But a great deal of our electricity is generated by oil-burning plants (Fig. 12-1). Every kilowatt saved is a few less barrels of oil. Cutting down on electrical consumption saves each of us directly in our electric bills. It also cuts down on imports by utilities, which saves us money indirectly, helps ease our dependence on other countries, and cuts down on inflation.

Fig. 12-1. Oil-burning power plants generate electricity in many areas of the United States. The stacks of this Long Island Lighting facility dominate the background of otherwise picturesque Port Jefferson harbor.

Some of you may question whether your personal electricity conservation will really cut down on oil imports. After all, you may say, my utility has no oil-generating plants at all. What good will it do if I cut back in my area? That's a good question. Well, besides saving you money, the other answer is that electric companies are actually tied together in gigantic "grids," the purpose of which is to eliminate the disastrous blackouts and brownouts that used to plague us, but which also enables utilities to share power production.

Some Ways to Conserve

There are numerous ways of cutting down energy consumption in the home. Many books have been written on this subject. Governmental agencies and utilities have numerous pamphlets and brochures on the same theme. All agree that one of the best and most cost-efficient methods is proper insulation.

Insulation

On the topic of insulation, we'll just say that no matter what type of fuel you use, insulation will pay back—in spades—whatever you invest. Even if you live in the Deep South and use little fuel for heating, you will find that insulation will save enormously in air-conditioning costs. This is a book about wiring and electricity, after all, and much as I'd like to wax enthusiastic about this topic, it is not directly pertinent. Suffice it to say that if your home is not insulated to the level shown in Figure 12-2, you're wasting precious fuel—and money.

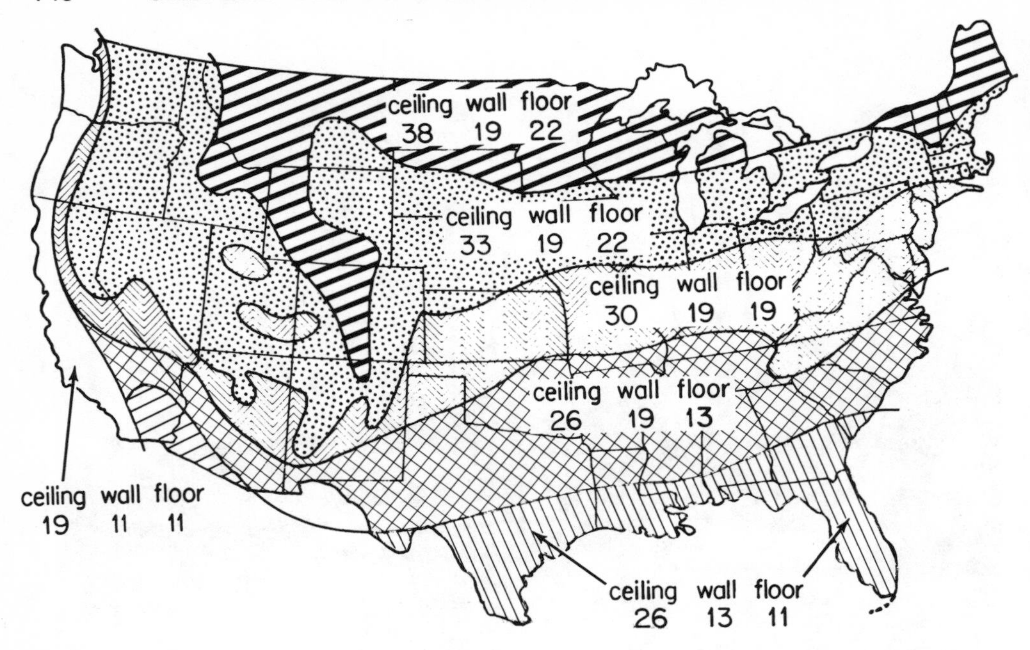

Fig. 12-2. Recommended insulation levels by R-factor in the United States. The seemingly high requirements in the South assume central air-conditioning.

Fig. 12-3. Caulking seals drafty cracks such as those between a chimney and siding. (Courtesy of Dow-Corning)

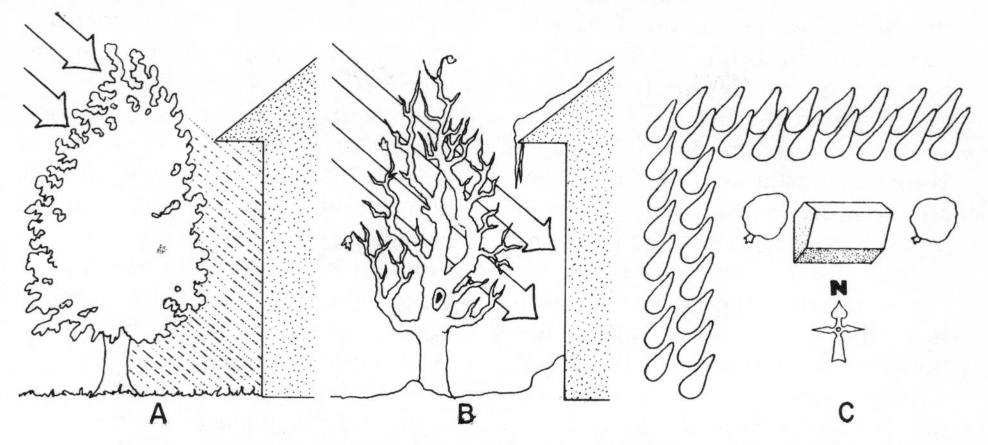

Fig. 12-4. One of the best and most satisfying energy-savers is a deciduous tree. (A) *In summer, the leaves shade the home from the hot rays of the sun.* (B) *In winter, the leaves fall off and allow the solar heat to penetrate.* (C) *Evergreen trees form a windbreak.*

Other general ways of preserving fuel are by proper caulking and weatherstripping (Fig. 12-3). Storm windows, wood stoves, solar heat, intelligent landscaping (Fig. 12-4), lowered thermostats in winter and raised ones in summer, heating-unit tune-ups, and numerous other energy-savers can be instituted by most homeowners. For more on this, see my book *Home Insulating* (Popular Science/ Harper & Row, 1978), among others.

As far as saving electricity *per se,* there are a great many things you can do, some of which save just a little, some a lot. But each step, if taken by all of us, could go a long way toward reducing our energy dependence. In the long run, you'll save money, too.

Cooling Energy Savers

Overcooling is expensive and wastes energy. Don't use or buy more cooling equipment capacity than you actually need.

- If you need central air-conditioning, select the smallest and least powerful system that will cool your home adequately. A larger unit than you need not only costs more to run but probably won't remove enough moisture from the air. Ask your dealer to help you determine how much cooling power you need for the space you have to cool and for the climate in which you live.
- Make sure the ducts in your air-conditioning system are properly insulated, especially those that pass through the attic or other uncooled spaces. This could save you almost 9 percent in cooling costs.
- If you don't really need central air-conditioning, consider using individual window or through-the-wall units in rooms that need cooling from time to time. Select the smallest and least powerful units for the rooms you need to cool. As a rule, these will cost less to buy and less to operate. Look for the highest EER (energy efficiency ratio).
- Install a whole-house ventilating fan in your attic or in an upstairs window to cool the house when it's cool outside, even if you have central air-conditioning. It will pay to use the fan rather than air-conditioning when the outside temperature is below 82 degrees. When windows in

the house are open, the fan pulls cool air through the house and exhausts warm air through the attic. (See page 161.)

- When you use air-conditioning, set your thermostat at 78° F., a reasonably comfortable and energy-efficient indoor temperature. The higher the setting and the less difference between indoor and outdoor temperature, the less outdoor hot air will flow into the building. If the 78° F. setting raises your home temperature 6 degrees (from 72° F. to 78° F., for example), you should save between 12 and 47 percent in cooling costs, depending on where you live.
- Don't set your thermostat at a colder setting than normal when you turn your air-conditioner on. It will not cool faster. It will cool to a lower temperature than you need and use more energy.
- Set the fan speed on high except in very humid weather. When it's humid, set the fan speed at low; you'll get less cooling but more moisture will be removed from the air.
- Clean or replace air-conditioning filters at least once a month. When the filter is dirty, the fan has to run longer to move the same amout of air, and this takes more electricity.
- Turn off your window air-conditioners when you leave a room for several hours. You'll use less energy cooling the room down later than if you had left the unit running. Attach a timer to re-start the unit shortly before your return.
- Consider using a fan with your window air-conditioner to spread the cooled air farther without greatly increasing your power use. But be sure the air-conditioner is strong enough to help cool the additional space.
- Don't place lamps or TV sets near your air-conditioning thermostat. Heat from these appliances is sensed by the thermostat and could cause the air-conditioner to run longer than necessary.
- Keep out daytime sun with vertical louvers or awnings on the outside of your windows, or draw draperies, blinds, and shades indoors. You can reduce heat gain from the sun by as much as 80 percent this easy way.
- Keep lights low or off. Electric lights generate heat and add to the load on your air-conditioner.
- Do your cooking and use other heat-generating appliances in the early morning and after dark hours whenever possible.
- Open the windows instead of using your air-conditioner or electric fan on cooler days and during cooler hours.
- Turn off the furnace pilot light in summer, but be sure it's reignited before you turn the furnace on again.
- Dress for the warmer indoor temperatures. Neat but casual clothes of lightweight open-weave fabrics are more comfortable. A woman will feel cooler in a lightweight skirt instead of slacks. A man will feel cooler in a short-sleeved shirt than in a long-sleeved shirt of the same weight fabric. Open-neck shirts and blouses feel cooler.
- Without air-conditioning, be sure to keep windows and outside doors closed during the hottest hours of the day. This will keep cooler air inside.

Hot Water Energy Savers

Heating water accounts for about 20 percent of all the energy we use in our homes. Don't waste it.

- Repair leaky faucets promptly. One drop a second can waste as much as 60 gallons of hot or cold water in a week.
- Do as much household cleaning as possible with cold water.

- Insulate your hot water storage tank and piping.
- Buy a water heater with thick insulation on the shell. While the initial cost may be more than one without this conservation feature, the savings in energy costs over the years will more than repay you.
- Add insulation around the water heater you now have if it's inadequately insulated, but be sure not to block off needed air vents. (Fig. 12-5). That would create a safety hazard, especially with oil and gas water heaters. When in doubt, get professional help. When properly done, you should save about 15 dollars a year in energy costs.
- Check the temperature on your water heater. Most water heaters are set for 140° F. or higher, but you may not need water that hot unless you have a dishwasher. A setting of 120° F. can provide adequate hot water for most families.

If you reduce the temperature from 140°F. (medium) to 120°F. (low), you could save over 18 percent of the energy you use at the higher setting. Even reducing the setting 10 degrees will save you more than 6 percent in water heating energy. If you are uncertain about the tank water temperature, draw some water from the heater through the faucet near the bottom and test it with a thermometer.

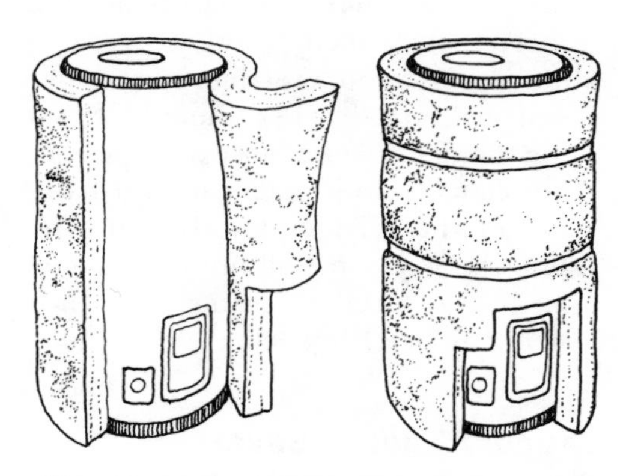

Fig. 12-5. An insulation blanket around your hot-water heater will save at least 15 dollars a year. Kits are available for this, but be careful not to cover up vents on oil or gas heaters.

- Don't let sediment build up in the bottom of your hot water heater; it lowers the heater's efficiency and wastes energy. About once a month, flush the sediment out by drawing several buckets of water from the tank through the water heater drain faucet.

Cooking Energy Savers

- Use cold water rather than hot to operate your food disposer. This saves the energy needed to heat the water, is recommended for the appliance, and aids in getting rid of grease. Grease solidifies in cold water and can be ground up and washed away.
- Install an aerator in your kitchen sink faucet. By reducing the amount of water in the flow, you use less hot water and save the energy that would have been required to heat it. The lower flow pressure is hardly noticeable.
- If you need to purchase a gas oven or range, look for one with an automatic (electronic) ignition system instead of pilot lights. You'll save an average of up to 47 percent of your gas use—41 percent in the oven and 53 percent on the top burners.
- If you have a gas stove, make sure the pilot light is burning efficiently—with a blue flame. A yellowish flame indicates an adjustment is needed.
- Never boil water in an open pan. Water will come to a boil faster and use less energy in a kettle or covered pan.
- Keep range-top burners and reflectors clean. They will reflect the heat better, and you will save energy.

- Match the size of pan to the heating element. More heat will get to the pan; less will be lost to surrounding air.
- If you cook with electricity, get in the habit of turning off the burners several minutes before the allotted cooking time. The heating element will stay hot long enough to finish the cooking for you without using more electricity.
- When using the oven, make the most of the heat from that single source. Cook as many foods as you can at one time. Prepare dishes that can be stored or frozen for later use or make all oven-cooked meals.
- Watch the clock or use a timer; don't continually open the oven door to check food. Every time you open the door heat escapes and your cooking takes more energy.
- Use small electric pans or ovens for small meals rather than the kitchen range or oven. They use less energy.
- Use pressure cookers and microwave ovens if you have them. They can save energy by reducing cooking time.

Dishwashing Energy Savers

The average dishwasher uses 14 gallons of hot water per load. Use its energy efficiently.

- Be sure your dishwasher is full, but not overloaded, when you turn it on.
- When buying a dishwasher, look for a model with air power and/or overnight dry settings. These features automatically turn off the dishwasher after the rinse cycle. This can save you up to one-third of your total dishwashing energy costs.
- Don't use the rinse/hold cycle on your machine. It uses 3 to 7 gallons of hot water each time you use it.
- Scrape dishes before loading them into the dishwasher so you won't have to rinse them.

Refrigerator/Freezer Energy Savers

- Don't keep your refrigerator or freezer too cold. Recommended temperatures: 38 to 40° F. for the fresh food compartment of the refrigerator; 5° F. for the freezer section. (However, if you have a separate freezer for long-term storage, it should be kept at 0° F).
- If you're buying a refrigerator, it's energy economical to buy one with a power-saver switch. Most refrigerators have heating elements in their walls or doors to prevent sweating on the outside. In most climates, the heating element does not need to be working all the time. The power-saver switch turns off the heating element. By using it, you could save about 16 percent in refrigerator energy costs.
- Consider buying refrigerators and freezers that have to be defrosted manually. Although they take more effort to defrost, these appliances use less energy than those that defrost automatically.
- Regularly defrost manual-defrost refrigerators and freezers. Frost build-up increases the amount of energy needed to keep the engine running. Never allow frost to build up more than ¼ inch.
- Make sure your refrigerator door seals are airtight. Test them by closing the door over a piece of paper or a dollar bill so it is half in and half out of the refrigerator. If you can pull the paper or bill out easily, the latch may need adjustment or the seal may need replacing.

Laundry Energy Savers

You can save considerable amounts of energy in the laundry through conservation of hot water and by using your automatic washers and dryers less often and more efficiently.

- Wash clothes in warm or cold water, rinse in cold. You'll save energy and money. Use hot water only if absolutely necessary.
- Fill washers (unless they have small-load attachments or variable water levels), but do not overload them.
- Use the suds saver if you have one. It will allow you to use one tubful of hot water for several loads.
- Don't use too much detergent. Follow the instructions on the box. Oversudsing makes your machine work harder and use more energy, and can cause machine damage.
- Presoak or use a soak cycle when washing heavily soiled garments. You'll avoid two washings and save energy.
- Fill clothes dryers but do not overload them.
- Keep the lint screen in the dryer clean. Remove lint after each load. Lint impedes the flow of air in the dryer and requires the machine to use more energy.
- Keep the outside exhaust of your clothes dryer clean. Check it regularly. A clogged exhaust lengthens the drying time and increases the amount of energy used.
- If your dryer has an automatic dry cycle, use it. Overdrying merely wastes energy.
- Dry your clothes in consecutive loads. Stop-and-start drying uses more energy because a lot goes into bringing the dryer up to the desired temperature each time you begin.
- Separate drying loads into heavy and lightweight items. Since the lighter ones take less drying time, the dryer doesn't have to be on as long for these loads.
- If drying the family wash takes more than one load, leave small, lightweight items until last. You may be able to dry them after you turn off the power, with heat retained by the machine from earlier loads.
- Save energy by using the old-fashioned clothesline. As a bonus, clothes dried outdoors often seem fresher and cleaner than those taken from a mechanical dryer.
- Remove clothes that will need ironing from the dryer while they still are damp. There's no point in wasting energy to dry them thoroughly if they only have to be dampened again.
- You can save ironing time and energy by "pressing" sheets and pillow cases on the warm top of your dryer. Fold them carefully, then smooth them out on the flat surface.
- Save energy needed for ironing by hanging clothes in the bathroom while you bathe or shower. The steam often removes the wrinkles for you.

Bathroom Energy Savers

- Take showers rather than tub baths, but limit your showering time and check the water flow if you want to save energy. It takes about 30 gallons of water to fill the average tub. A shower with a flow of 4 gallons of water a minute uses only 20 gallons in 5 minutes. Assuming you use half hot and half cold water, you would save about 5 gallons of hot water every time you substitute a shower for a bath. Thus, if you substituted just one shower for one bath per day, you would save almost 2,000 gallons of hot water in a year. (But watch out for teenagers, who seem to revel in 20-minute showers, and who would save money by using the tub.)
- Consider installing a flow restrictor in the pipe at the showerhead. These inexpensive, easy-to-install devices restrict the flow of water to an adequate 3 to 4 gallons per minute. (Some utilities give these away.) This can save considerable amounts of hot water and the

energy used to produce them over a year's time. For example, reducing the flow from 8 to 3 gallons a minute would save the average family about 24 dollars a year, per U.S. Government studies.

Lighting Energy Savers

More than 16 percent of the electricity we use in our homes goes into lighting. Most Americans overlight their homes, so lowering lighting levels is an easy conservation measure.

- Light-zone your home and save electricity. Concentrate lighting in reading and working areas and where it's needed for safety (stairwells, for example). Reduce lighting in other areas, but avoid very sharp contrasts.
- To reduce overall lighting in non-working spaces, remove one bulb out of three in mutliple light fixtures and replace it with a burned-out bulb for safety. Replace other general-lighting bulbs throughout the house with bulbs of the next lower wattage. (But see preceding chapter.)
- Consider installing solid-state dimmers or hi-low switches when replacing light switches. They make it easy to reduce lighting intensity in a room and thus save energy.
- Use one large bulb instead of several small ones in areas where bright light is needed.
- Use long-life incandescent lamps only in hard-to-reach places. They are less energy efficient than ordinary bulbs.
- Need new lamps? Consider the advantages of those with three-way switches. They make it easy to keep lighting levels low when intense light is not necessary, and that saves electricity. Use the high switch only for reading or other activities that require brighter light.
- Always turn three-way bulbs down to the lowest lighting level when watching television. You'll reduce the glare and use less energy.
- Use low wattage night-light bulbs. These now come in 4-watt as well as 7-watt sizes. The 4-watt bulb with a clear finish is almost as bright as the 7-watt bulb but uses about half as much energy.
- Try 50-watt reflector floodlights in directional lamps (such as pole or spot lamps). These flood lights provide about the same amount of light as the standard 100-watt bulbs but at half the wattage.
- Try 25-watt reflector flood bulbs in high-intensity portable lamps. They provide about the same amount of light but use less energy than the 40-watt bulbs that normally come with these lamps.
- Use fluorescent lights whenever you can; they give out more lumens per watt (Fig. 12-6). For example, a 40-watt fluorescent lamp gives off 80 lumens per watt and a 60-watt incandescent gives off only 14.7 lumens per watt. The 40-watt fluorescent lamp would save about 140 watts of electricity over a 7-hour period. These savings, over a period of time, could more than pay for the fixtures you would need to use fluorescent lighting.
- Consider fluorescent lighting for the kitchen sink, and countertop areas. These lights set under kitchen cabinets or over countertops are pleasant and energy efficient. New, plug-in types are easy to install.
- Fluorescent lighting also is effective for makeup and grooming areas. Use 20-watt deluxe warm white lamps for these areas.
- Keep all lamps and lighting fixtures clean. Dirt absorbs light.
- You can save on lighting energy through decorating. Remember, light colors for walls, rugs, draperies, and upholstery reflect light and, therefore, reduce the amount of artificial light required.

Fig. 12-6. Fluorescent lights are cleverly used in this attractive basement room. The lights in the suspended ceiling provide general illumination. There are also fluorescents in the boxes behind the handsome stained-glass "windows." (Courtesy of Armstrong Cork)

- Use outdoor lights only when they are needed. One way to make sure they're off during the daylight hours is to put them on a photocell unit or timer that will turn them off automatically.

Appliance Energy Savers

About 8 percent of all the energy used in the United States goes into running electrical home appliances, so appliance use and selection can make a considerable difference in home utility costs. Buying energy-efficient appliances may cost a bit more initially, but that expense is more than made up by reduced operating costs over the lifetime of the appliance.

- Don't leave your appliances running when they're not in use. Remember to turn off your radio, TV, or record player when you leave the room.
- Keep appliances in good working order so they will last longer, work more efficiently, and use less energy.
- When buying appliances, comparison shop. Compare energy use information and operating costs of

similar models by the same and different manufacturers. The retailer should be able to help you find the wattage of the appliance. With that information, and the chart on page 155, you should be able to figure out how much it will cost you to run the appliance you choose. Newer appliances should show operating costs on the label.

- Before buying new appliances with special features, find out how much energy they use compared with other, perhaps less convenient, models. A frost-free refrigerator, for example, uses more energy than one you have to defrost manually. It also costs more to purchase. The energy and dollars you can save with a manual-defrost model may be worth giving up the convenience.
- Use appliances wisely; use the one that takes the least amount of energy for the job. For example, toasting bread in the oven uses three times more energy than toasting it in a toaster.
- Don't use energy-consuming special features on your appliances if you have an alternative. For example, don't use the instant-on feature on your TV set. Instant-on sets, especially tube-types, use energy even when the screen is dark. Use the vacation switch, if you have one, to eliminate this waste; plug the set into an outlet that is controlled by a wall switch; or have your TV service man install an additional on-off switch on the set itself or in the cord to the wall outlet.

Building or Buying an Energy-Efficient Home

Energy-wasting mistakes can be avoided if you consider climate, local building codes, and energy-efficient construction when you build or buy a home. In either case, the following energy conservation ideas should help you keep down home utility bills.

- Consider a square floor plan. It usually is more energy efficient than a rectangular plan.
- Insulate walls and roof to the highest specifications recommended for your area (see page 145).
- Insulate floors, too, especially those over crawl spaces, cold basements, and garages.
- If the base of a house is exposed, as in the case of a mobile home, build a "skirt" around it.
- Install louvered panels or wind-powered roof ventilators rather than motor-driven fans when possible.
- Consider solar heat gain when you plan your window locations. In cool climates, install fewer windows in the north wall because there's little solar heat gain there in winter. In warm climates, put the largest number of windows in the north and east walls to reduce heating from the sun.
- Install windows you can open so you can use natural or fan-forced ventilation in moderate weather.
- Use double-pane or triple-pane glass throughout the house. Windows with double-pane heat-reflecting or heat-absorbing glass provide additional energy savings, especially in south and west exposures.
- Place your refrigerator in the coolest part of the kitchen, well away from the range and oven.
- Install the water heater as close as possible to areas of major use to minimize heat loss through the pipes; insulate the pipes.
- If you live in a warm climate, remember that light-colored roofing can help keep houses cooler.

Annual Energy Requirements of Electric Appliances

	Estimated Kilowatt Hours Used Annually
Major appliances	
Air-conditioner (room) (Based on 1,000 hours of operation per year)	860
Clothes dryer	993
Dishwasher (including energy used to heat water)	2,100
Dishwasher only	363
Freezer (16 cu. ft.)	1,190
Freezer—frostless (16.5 cu. ft.)	1,820
Range with oven	700
Range with self-cleaning oven	730
Refrigerator (12 cu. ft.)	728
Refrigerator—frostless (12 cu. ft.)	1,217
Refrigerator/freezer (12.5 cu. ft.)	1,500
Refrigerator/freezer—frostless (17.5 cu. ft.)	2,250
Washing machine (including energy used to heat water)	2,500
Washing machine only	103
Water heater	4,811
Kitchen appliances	
Blender	15
Broiler	100
Carving knife	8
Coffee maker	140
Deep fryer	83
Egg cooker	14
Frying pan	186
Hot plate	90
Mixer	13
Oven, microwave only	190
Roaster	205
Sandwich grill	33
Toaster	39
Trash compactor	50
Waffle iron	22
Waste disposer	30
Heating and cooling	
Electric blanket	147
Dehumidifier	377
Fan (attic)	291
Fan (circulating)	43
Fan (rollaway)	138
Fan (window)	170
Heater (portable)	176
Heating pad	10
Humidifier	163
Laundry	
Iron (hand)	144
Health and beauty	
Germicidal lamp	141
Hair dryer	14
Heat lamp (infrared)	13
Shaver	1.8
Sun lamp	16
Toothbrush	0.5
Vibrator	2
Home entertainment	
Radio	86
Radio/record player	109
Television	
Black and white	120
Color	440
Housewares	
Clock	17
Floor polisher	15
Sewing machine	11
Vacuum cleaner	46

Note: When using these figures for projections, such factors as the size of the specific appliance, the geographic area of use, and individual use should be taken into consideration.

Source: Edison Electric Institute.

13

Electrical Projects

This chapter contains a miscellany of things which are powered by electricity that you can make for home and family. Some are simple and require little or no knowledge of electricity or wiring. Others are more complicated and demand more *savoir faire*. Some, like the lamps, can be made from common household items and some hardware. Others must be purchased from various manufacturers.

Attractive Lamps from Household Items

A table lamp is basically a simple device. As explained in Chapter 5, there are only a few necessary parts, such as a light bulb, a socket, a switch, and a cord. Other hardware parts are necessary, however, to perform such mechanical functions as holding up the shade, containing the wires, and attaching the electrical parts to the lamp base.

If lamps consisted of only simple electrical parts, however, there would be no need for designers, lamp-factory workers, shade manufacturers, how-to writers, and assorted other people. The aesthetic aspect of lamp-making is what sells lamps. When making your own, the creative possibilities are legion, with the only limitation your imagination.

Creative Selection of Materials

Lamps have been made out of wicker baskets, flower pots, jars of all sorts, children's blocks, wine bottles, fishbowls, gumball machines, statues, driftwood—you name it. Plain jars and bottles can be filled with colored sand or beach pebbles, dried flowers, leftover yarn, rice, herbs, almost anything colorful. Plywood or plastic foam can be cut to a desired shape and covered with fabric, wallpaper, tiles, mirror squares, colored rulers, shells, or anything else that suits your fancy.

Depending on the type of bottom on the base, and its weight, you may have to fill the base with sand or gravel to prevent its being tipsy (especially a former liquor bottle). If you've cut a hole in the bottom of the base to allow for the center rod, enclose the sand in small plastic bags to prevent spillage.

Technique

To make a professional-looking lamp, run the wiring up from the bottom of the base through a threaded rod as shown in Figure 13-1. You can make life easier for yourself, however, if you buy a kit like the one shown in Figure 13-2. Since the cord runs directly from the socket, there is no need to extend the wires through the base. This may be necessary when working with glass, for example, which is dif-

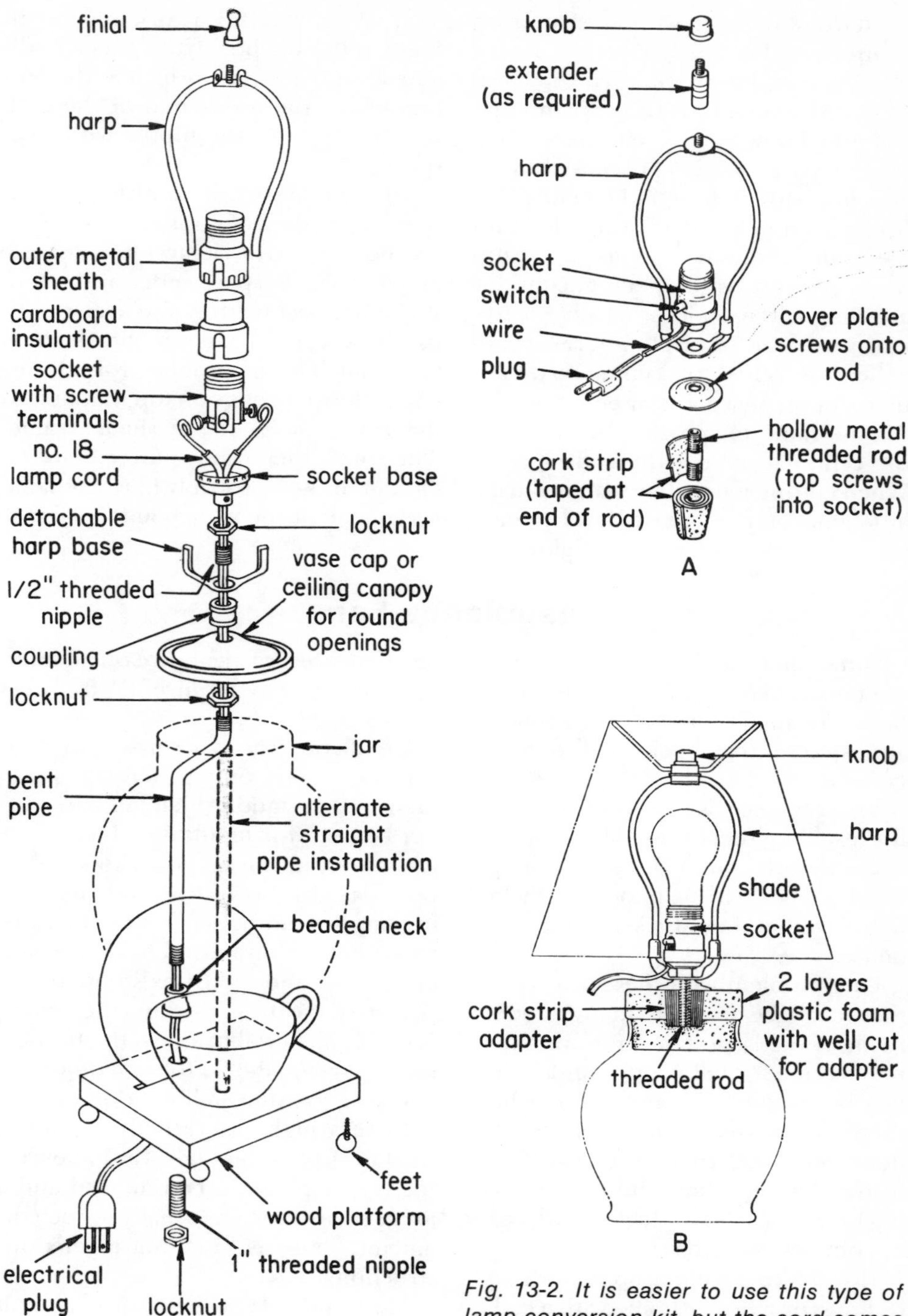

Fig. 13-1. A really handsome lamp can be created out of numerous materials using the parts shown and drilling through the bottom of the jar or other base.

Fig. 13-2. It is easier to use this type of lamp conversion kit, but the cord comes out of the top and is more visible. In some situations, though, where the back can be hidden, the cord will never be seen. (A) Lamp fixture (B) Lamp assembly.

ficult to cut through without shattering completely. The cord, of course, is visible that way, but if you turn the lamp so that the cord is at the rear, no one will ever know you've "cheated" a bit.

Most parts can be purchased at an electrical supply house. The chief difficulty is often fitting the threaded nipple securely into the opening at the top of base. For a large opening, consider a ceiling canopy piece, which is ordinarily used decoratively where a chandelier is attached to a ceiling box. Plastic foam or plywood can be cut to size for the same purpose, then drilled in the center to accept the rod or cork. When cutting a hole in foam, make it a little smaller than needed. The foam will compress and make a tighter fit. To make a hole for a cork adapter in foam, taper the hole from 1 inch in diameter at the top to ¾ inch at the bottom. Tape the inside end of the cork strip securely to the nipple with electrician's tape.

You may have a little difficulty getting the shade you want to fit correctly on the harp. If it sits too low, for example, use a harp extender at the top. A clamp attachment is also available to use instead of a harp, if the shade is too small. This is a double, round wire clip with two prongs on top that fit into the center hole in the shade frame. The shade then clamps directly to the light bulb. You probably have the same arrangement for your boudoir lamps.

Casablanca Fans

Some manufacturers call these paddle fans. I like to call them Casablanca fans. The name connotes the romantic atmosphere that such fans seem to generate.

But romance isn't the only reason for installing these fans. They also cool a room with minimal use of energy. Most come with optional attached lights, so that they also serve as a chandelier.

One problem with these fans is that they can't be used in many modern homes. The large fan arms are a definite hazard for tall people, and should not be installed in any room where there is less than 7 feet between the floor and the trailing edge of the blades. For most fans, this means a ceiling height of at least 8 feet—hardly easy to come by these days.

If you have an older home with high ceilings, or a newer home with a cathedral ceiling, a Casablanca fan can be a definite asset for both looks and comfort. In addition to height requirements, there should also be a minimum of 2 feet of clearance on all sides. Larger fans may require 2½ to 3 feet of clearance.

Installation varies according to manufacturer, but one general requirement is that standard hanger bars cannot be used for mounting. These fans are heavy, in the neighborhood of 50 pounds, and the 4-inch octagonal ceiling box should be securely screwed to a joist bottom or a wood header nailed between the joists. In existing rooms, that may mean tearing up the ceiling finishing material to attach the header. (See pages 98–99 for methods of patching gypsum wallboard.)

In very high and cathedral ceilings, the fan may be *too* high. In that event, the center pipe can be removed and a longer one inserted (usually ½-inch diameter, but see manufacturer's instructions).

Most models come with a pull switch, which can be converted to a regular or variable-speed wall switch. Some models can also be installed "swag" style and plugged in like a stan-

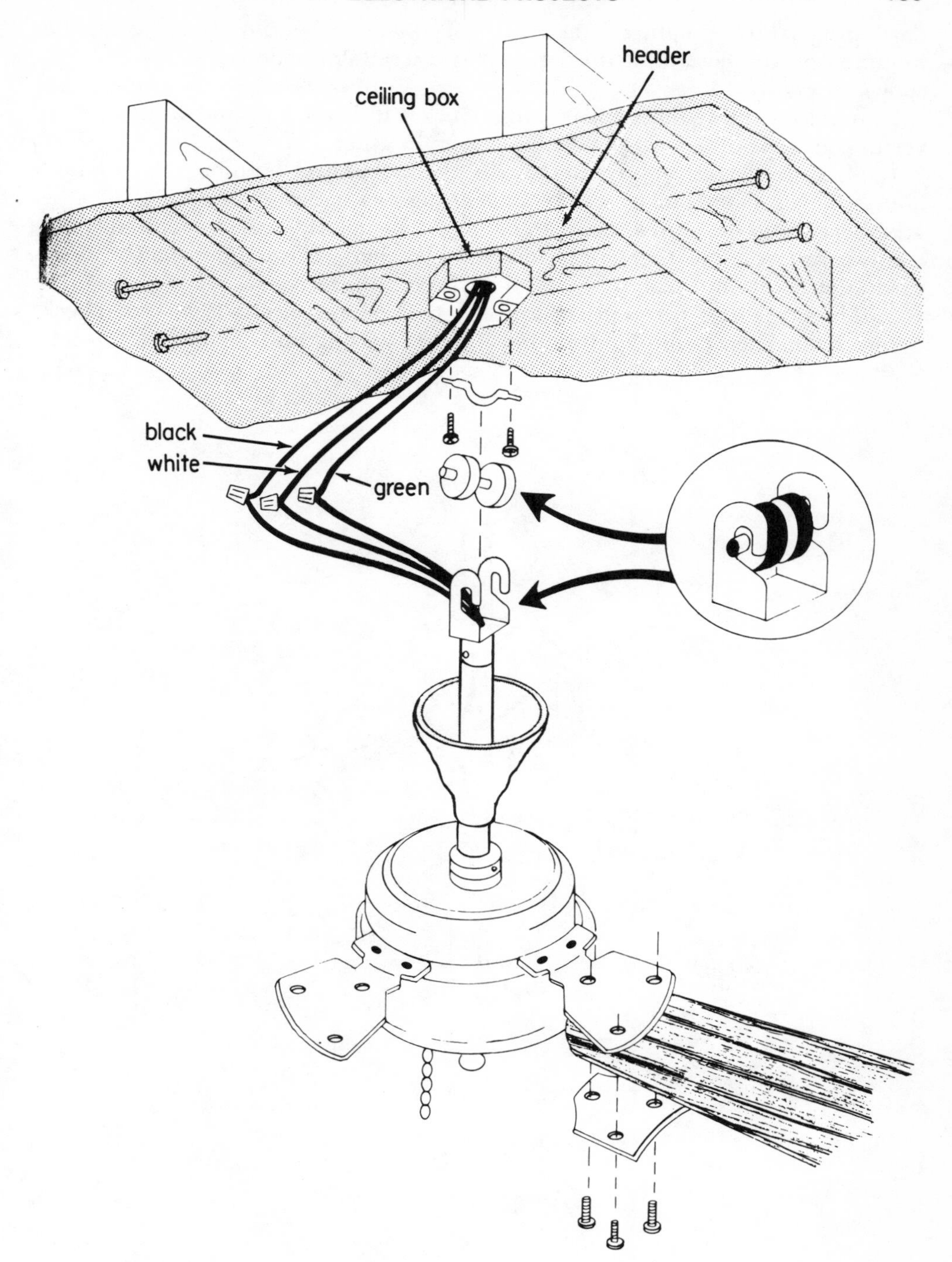

Fig. 13-3. The workings of a typical Casablanca fan. Note the header placed between joists to give a sturdy support for the heavy fan.

dard swag lamp. A light can also be mounted on the bottom of the fan as shown in Figure 13-4.

Figure 13-4 shows the steps in converting one type of fan to a paddle fan.

1. Shut off electric supply. Remove bottom cover from fan by removing knob screw and decorative plate, then the two screws. Save the screws, discard knob screw and plate.

2. Thread wires through bottom cover of fan, then fasten light mounting plate to bottom cover, using the two screws provided.

3. Connect two wires from lamp socket to black pair and white pair in fan housing.

4. Reconnect bottom cover of fan, using two screws.

5. Install 100-watt (maximum light bulb, then secure lampshade, using three thumbscrews.

6. Restore electricity, and check light by using pull chain.

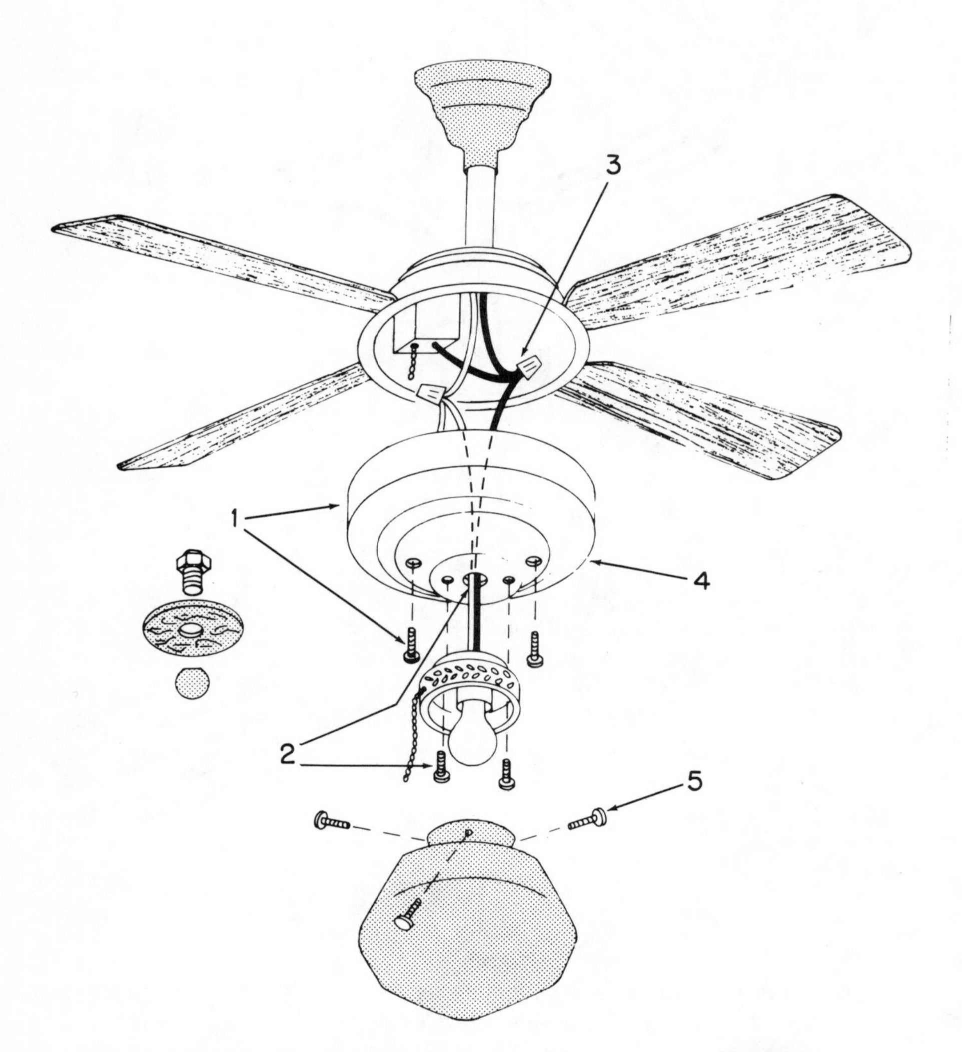

Fig. 13-4. Many Casablanca fans can also serve as ceiling fixtures. Drawing shows the steps in converting one type of fan.

Exhaust Fans

There are fans and fans, all of which have the same general purpose of moving air around. The reasons for moving the air may differ, however. A kitchen or bathroom fan is used to remove odors, while a window fan is used to cool a room. The Casablanca fan just discussed is also a mood-maker. A real fan *aficianado* (a fan fan?) knows that the intelligent use of fans can provide a lot of cooling at considerably lower cost than air-conditioning. There are two general ways to use permanently installed fans to provide this type of cooling.

The Whole-House Fan

This is the most efficient form of cooling. It is relatively expensive, requires a bit of work to install, and can be noisy, but the benefits are enormous. Installed on the attic floor, it pushes warm air out of the attic, while at the same time drawing cool air through the windows of the rooms below. Some are capable of drawing up to 7,000 cubic feet of air per minute.

Although whole-house fans can be installed vertically, they are more effective when mounted on the ceiling of the top floor (on the attic floor). Automatic shutters open whenever the fan goes on, and stay closed during cooler days.

Fig. 13-5. Attic fans draw cool outside air into the house and expel warm air from inside.

Researchers say that the energy savings from a whole-house fan can run from 10 to 66 percent as opposed to air-conditioning, and depending on the type of house and location. They won't help too much, however, when the temperature doesn't drop 20 to 30 degrees at night. The fan's breeze also helps evaporate water and sweat in high humidity, thus making you feel a lot cooler.

In general, whole-house fans are a good investment when the *average* temperature is 75 degrees in July and August. Remember, that's an average, and even though it may seem a lot hotter than that, you'll probably be surprised that the average is lower than that in most temperate zones.

Size

When buying a whole-house fan, be sure to get the right size for your home. The proper way to do this is to calculate the cubic feet of all the rooms you'll be cooling. You can ignore closets, all storage areas, and unused rooms. If you will not be cooling other rooms, such as a dining room, most of the time, and it can be closed off, you can ignore those also.

When you have added up the space in all the rooms that will be cooled or cannot be closed off, you have the number of cubic feet that must be moved by the fan. In Sunbelt areas, one air exchange per minute is advisable. More temperature zones may need only one exchange every two minutes. Fans are rated at air exchanges per cubic feet. Thus, a home with a volume of 7,000 cubic feet in the South would need a large fan with a rating of 7,000 cfm (cubic feet per minute). The same home in the North would need one only half as large.

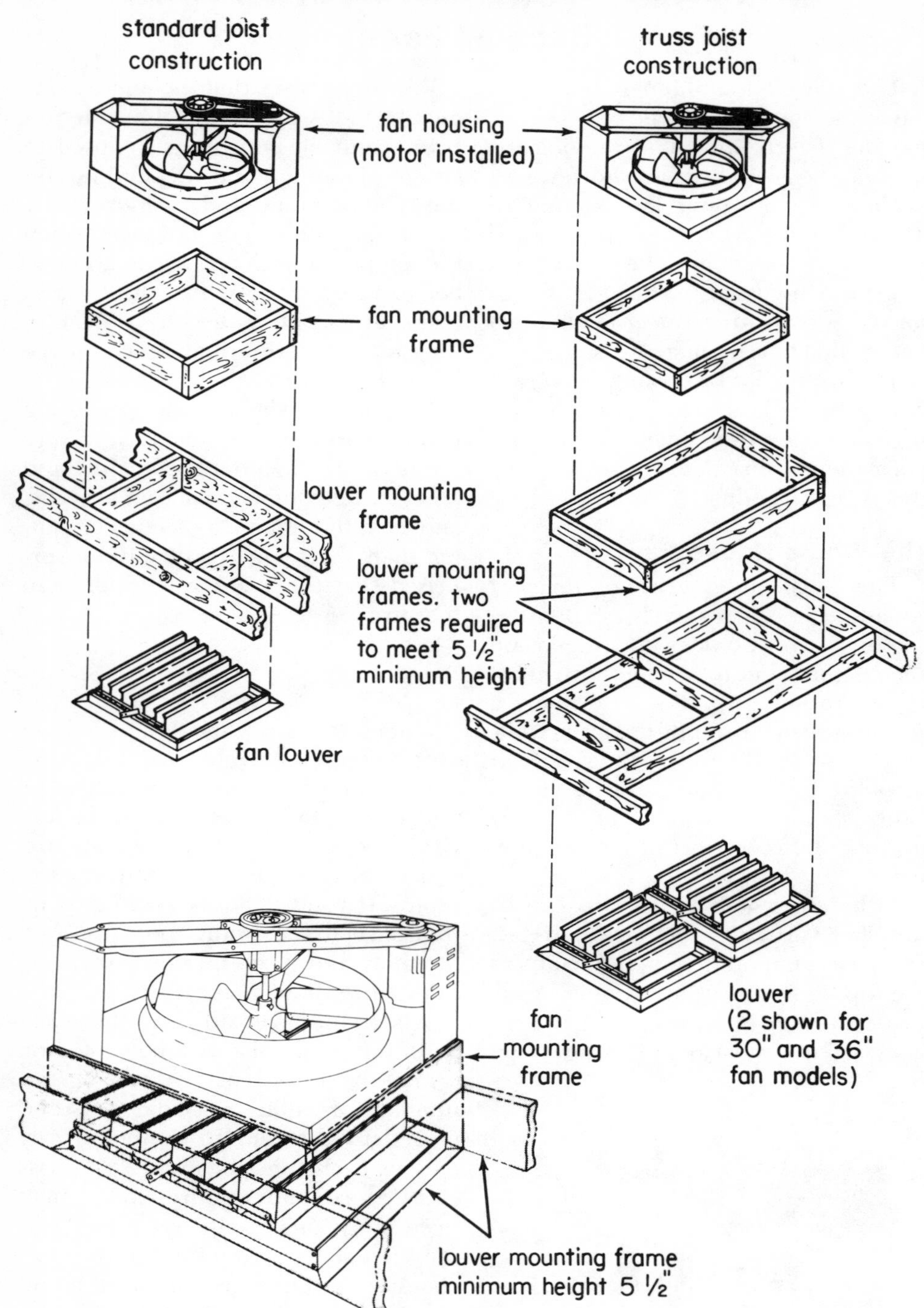

Fig. 13-6. When joists must be cut, as they undoubtedly will be for any large fan, a separate frame must be erected around the fan. With truss-type roofs, double louvers are needed, because joists cannot be cut. (Top left) Standard joist construction. (Top right) Truss-joist construction. (At bottom) Louver mounting frame.

Placement

Another important consideration is placement of the fan. It should be centrally located with plenty of room overhead. Since the smallest unit is about 2 feet wide, you will probably have to cut through some framing on installation. So try to locate the unit so that a minimum of joist removal is necessary. If you have a trussed roof (with diagonal framing members between the rafters and the ceiling joists), buy a unit with double shutters, so you won't have to cut any framing. You can't cut truss framing without bringing down the roof—literally. (See Figure 13-6 for truss-type installation.)

When you've located the fan optimally, mark holes at the four corners of the unit and drill holes at each. Going back downstairs, draw lines between the holes with a straightedge. Cut along the lines with a utility or razor knife for gypsum wallboard, or a saber saw for wood.

Where joists must be cut, double up the framing on each side of the hole as shown in Figure 13-6 and complete any other framing using the same-sized lumber as the joists. Some installations require a 1x4 plate on top of the frame. Since this type of fan is quite heavy, special mounting is usually not required. The fan simply rests on a rubber gasket. That same weight factor, incidentally, will also require that you have some assistance in lifting the fan to the attic.

After the fan is set in place, the shutters are installed in the space below and the wiring is hooked up. This is ordinarily done through a thermostat switch on a separate circuit. Follow manufacturer's instructions for this phase of the installation.

You should be aware of the fact that the large volume of air that is expelled by the fan requires a vent area in the attic at least as large as the fan itself. You will probably have to enlarge the vents in your attic, supplanting the present louvers with a shutter-type device similar to the ones on the fan housing. Check manufacturer's recommendations for this.

Attic Space Ventilator

Although an attic fan looks like a whole-house fan, and follows many of the same principles, its purpose is quite different. The attic ventilating fan is designed to cool only the attic itself.

Cooling the attic alone may seem like a rather trivial step toward home comfort, but it can play a large part in keeping your house comfortable. That's because the attic retains a great deal of the heat that is built up in the house during the day, then releases that heat downward during the cooler night hours. It is not unusual, during a 90-degree-day, for the attic to heat up to 130 degrees. At 95 degrees, the attic can go up to 150° F. (If you're skeptical, take a trip up there the next hot day. Bring a thermometer.)

Exhausting that hot air from the attic will keep it from infiltrating below. Adequate attic insulation also helps, but a ventilating fan makes the insulation's job that much easier.

Placement

Attic fans can be placed on the roof or in an attic sidewall or gable (Fig. 13-7). In either case, it should be installed as high up as possible. Louvers open and close automatically as needed. Roof-top models do not require cutting of the rafters (a dangerous practice, if they did), but a vertical fan will probably require cutting into a stud. For side-wall fans, construct a 2x4 frame to fit around the fan housing, as shown in Figure 13-8. The fan is then mounted to the exterior surface of the building.

To install a rooftop fan, drill a hole from the attic through the center of

Fig. 13-7. Roof fans are highly efficient in removing hot air from attic. Properly designed and installed, as this one is, they keep rain out of the attic, too.

the roof sheathing between two rafters (Fig. 13-9A). With the drilled hole as the center, draw a circular line the same diameter as the fan housing. Use a string compass or piece of wood as shown in Figure 13-9B, with a radius half the diameter of the circle (for a 14½-inch opening, for example, the radius of the circle is 7¼ inches).

Cut out the circle with a saber saw. (Fig. 13-9C). Attach to the roof as directed by the manufacturer and caulk around the edges (Fig. 13-9D). Since it is self-contained, no other installation is necessary, except for the electrical work. The thermostat-switch is located along a rafter somewhere above the fan, then hooked up to its own circuit.

It is important that any attic ventilating fan have sufficient inlet space. For each 300 cfm rating of the fan, a square foot of inlet space is required. Thus, a 1,400 cfm fan needs almost 5

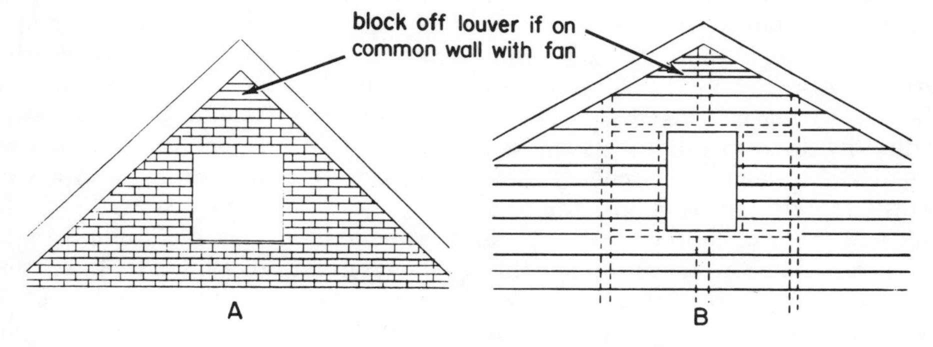

Fig. 13-8. How to make an opening for attic sidewall fan (A) in brick and (B) in wood siding. Install new framing all around opening for wood.

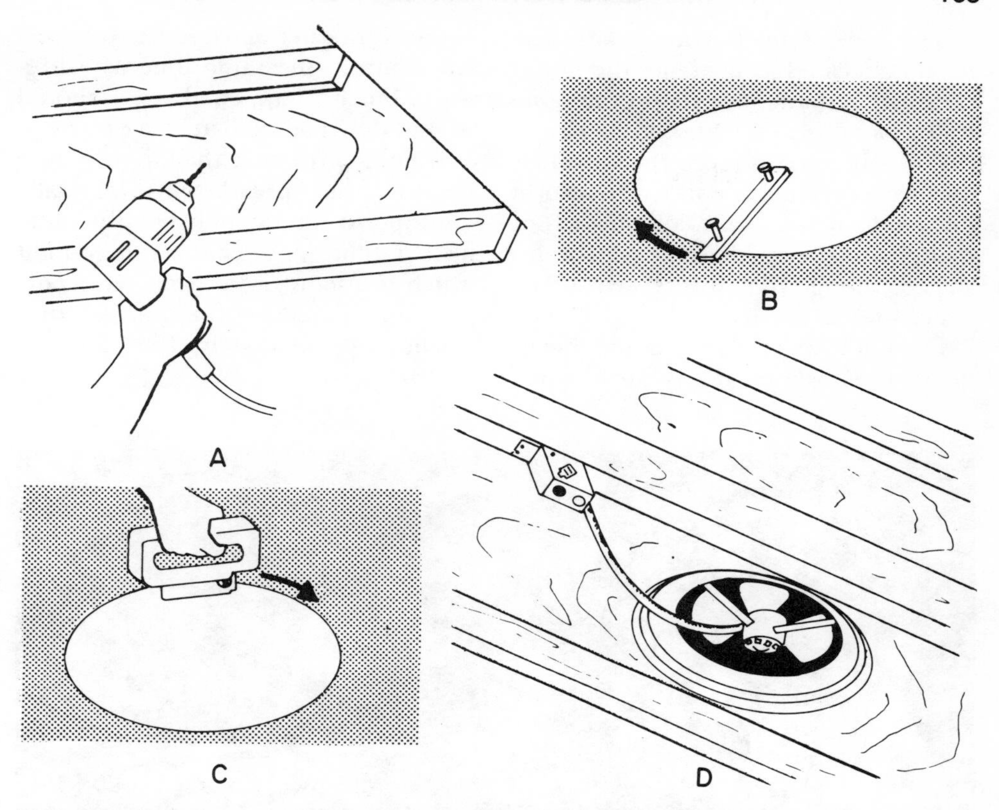

Fig. 13-9. (A) First step in installing a rooftop fan is to drill a hole from below, centered between rafters. (B) With a string or wood compass (shown), draw a circle corresponding to the diameter of the fan. (C) Cut out a hole for the fan with a saber or keyhole saw. (D) Install the fan per manufacturer's directions, making sure that the shingles overlap the flanges, then install the thermostat-switch. It should be located on the rafter somewhere above the fan.

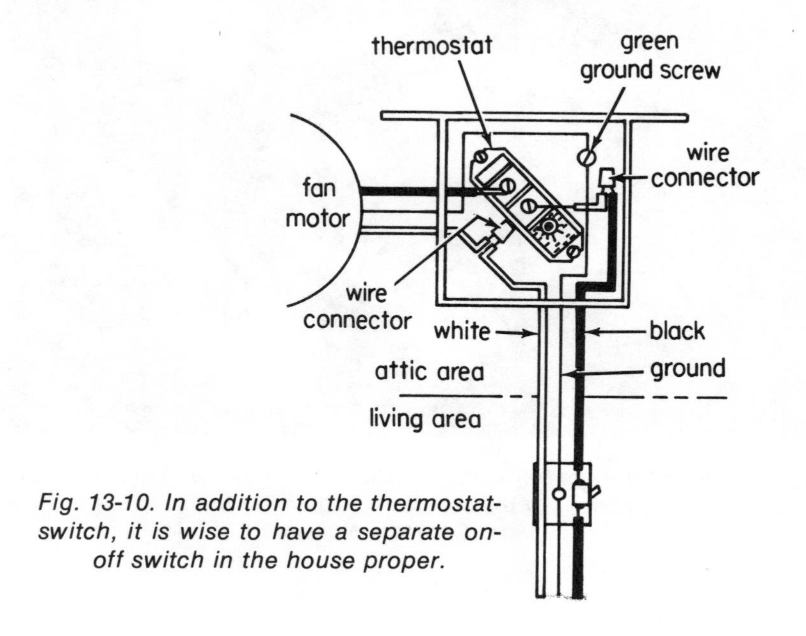

Fig. 13-10. In addition to the thermostat-switch, it is wise to have a separate on-off switch in the house proper.

square feet of inlet space. The inlets must be located as far from the fan as possible, and any louvers on the same wall as a vertical fan should be closed off. The best location for the louvers is on the opposite sidewall for a vertical fan, and on both sidewalls for a roof model. Additional louvers can be installed along the soffits (eaves or overhang) as needed.

IMPORTANT: Fans should never be operated when a fire is burning in a house. It doesn't make sense to have a fan going at the same time as a fireplace, but it's happened. Fans should be shut down immediately in case of an unfriendly fire, since the fan will cause a more rapid spread. Since manually shutting off the fan may be difficult at such a time, a disconnect high-limit switch is also advised, which shuts off the fan whenever the temperature reaches approximately 160° F. (Fig. 13-10).

Fig. 13-11. Compact storage trays designed to fit into most kitchen cabinets hold all of the nine different appliances of the compact food center that run off the single power unit. (Courtesy NuTone)

A Compact Food Service Center

Modern food processors are all the rage these days, and they do perform an admirable array of kitchen tasks. They do them well, too. But here's a kitchen "gadget" that can be a food processor and a lot more besides. It's an ideal homemaker's tool, particularly if your kitchen is small and lacks sufficient counter and/or storage space for the many appliances the modern cook is heir to (Fig. 13-11).

The basic premise is a simple one. You install one versatile multispeed power unit into the counter, then use it to run a wide variety of appliances such as blender, mixer, juicer, ice crusher, knife sharpener, meat grinder, can opener and slicer-shredder—in addition to a standard food processor.

If you bought all of the attachments at once, this could be an expensive item, although you'd still save money over buying each appliance individually. A beginning homemaker on a budget, in a small home or condominium, can buy just the power unit and a few attachments at first, then add on as the family grows. (An apart-

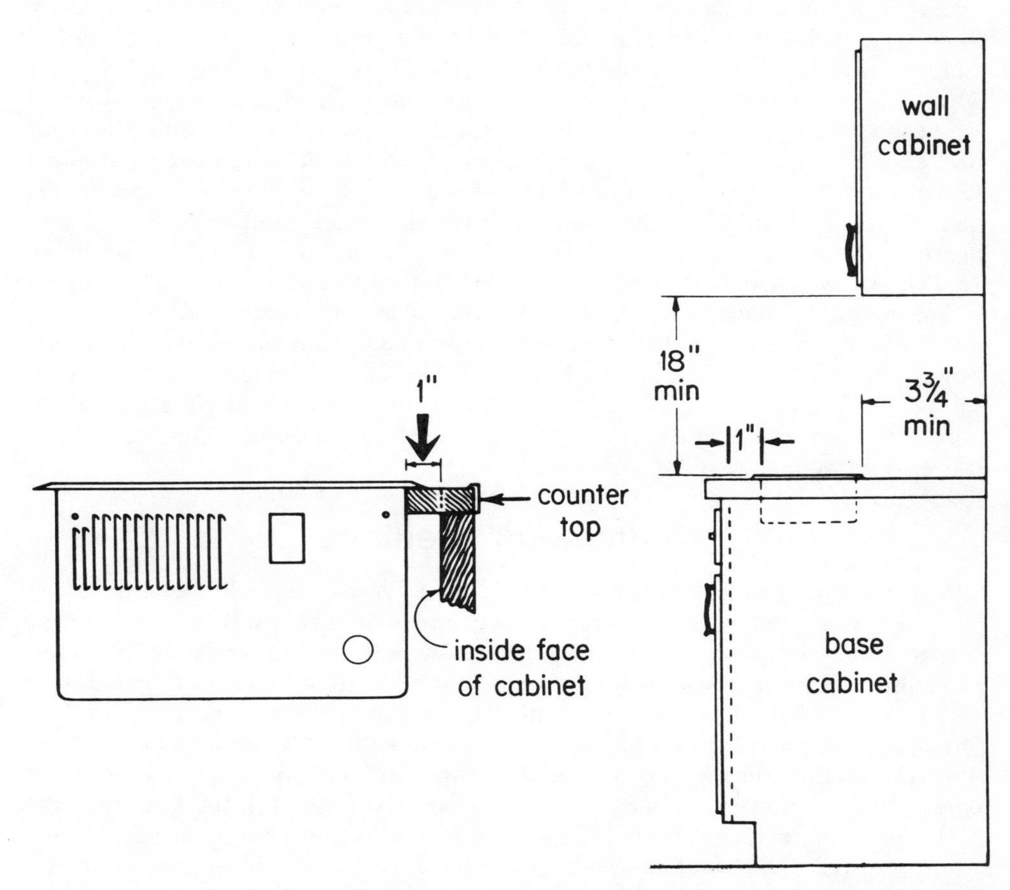

Fig. 13-12. The power unit should be installed 1 inch from the front of the cabinet, and have at least 18 inches clearance from the top cabinet. A minimum of 3¾ inches is required between walls and all sides. Unit also needs at least 6¾ inches clearance underneath, measured from the top of the counter.

ment dweller, with the typical small kitchen, could benefit most, but will probably need permission from the owner, and have to leave the power unit behind if and when she moves.)

Installation is quite simple. Most of the attachments operate at low wattages, but some go up to 400 watts, and the processor uses 650 watts. Check out your circuitry, and remember that you will be replacing other appliances with this unit. You should be able to add the power unit to most circuits without taxing the circuit capacity. It is easy to install.

The control dial is self-contained, so no switches are necessary. All you have to do is run a line to power unit and attach the wires as usual (black to black, white to white, and ground to ground).

Location is important. The power unit should be placed in a convenient spot where there is a minimum of 18 inches overhead and 6¾ inches underneath from the top of the counter (Fig. 13-12). Allow 3¾ inches between adjacent walls and the unit. There should also be at least an inch between the power unit and the inside of the base cabinet.

A template is supplied with the power unit. When an appropriate spot is found, hold the template in place on the counter and draw around the edges. Make a cutout with a keyhole or saber saw (Fig. 13-13 A). Use a fine-tooth blade in laminated plastic to avoid chipping.

Make the wiring connections as discussed (see Figure 13-13 B), then remove the top plate from the power unit by squeezing the tension springs as shown in Figure 13-13 C. Place unit into the cutout with the junction box in front and the power spindle at the rear. Secure the unit to the countertop by turning the mounting screw (Fig. 13-13 D) clockwise until the clamp bracket is snugly against the underside of the counter. Do not overtighten.

Position the top plate over the power unit and squeeze the tension springs together. Insert them into the slots provided. Don't push plate all the way down yet. First align the top gasket over the motor well. When plate and gasket are in line, push the plate down until it snaps flush against the countertop. Place the control dial onto the switch shaft and push down. (Fig. 13-13 E). Set the motor well cover in place. Remove the well cover when using the various attachments.

"Luminous" Ceilings

A "luminous" ceiling has several advantages. For one, it uses energy-efficient fluorescent lights. With proper planning, the lighted area will receive softer, more even, more adequate illumination. Acoustical panels or tiles installed in conjunction with translucent panels also cut down on noise.

If you have relatively high ceilings, you can install a lower ceiling, perhaps using insulating panels, which save precious fuel in two ways. There will be fewer cubic feet to heat or cool, and the insulating panels will help eliminate heat loss (top floors only).

But you aren't limited to those rooms where you will be lowering the ceiling. If you already have a suspended ceiling, you can install new lights above, replacing the standard panels with translucent, plastic types. And it isn't only the kitchen that can be so utilized (although the kitchen is the primary location for this type of ceiling). It is also possible to install fluorescent fixtures and translucent panels in living rooms or other areas that have suspended ceilings, as shown in Figure 13-14.

Fluorescent lights can be mounted

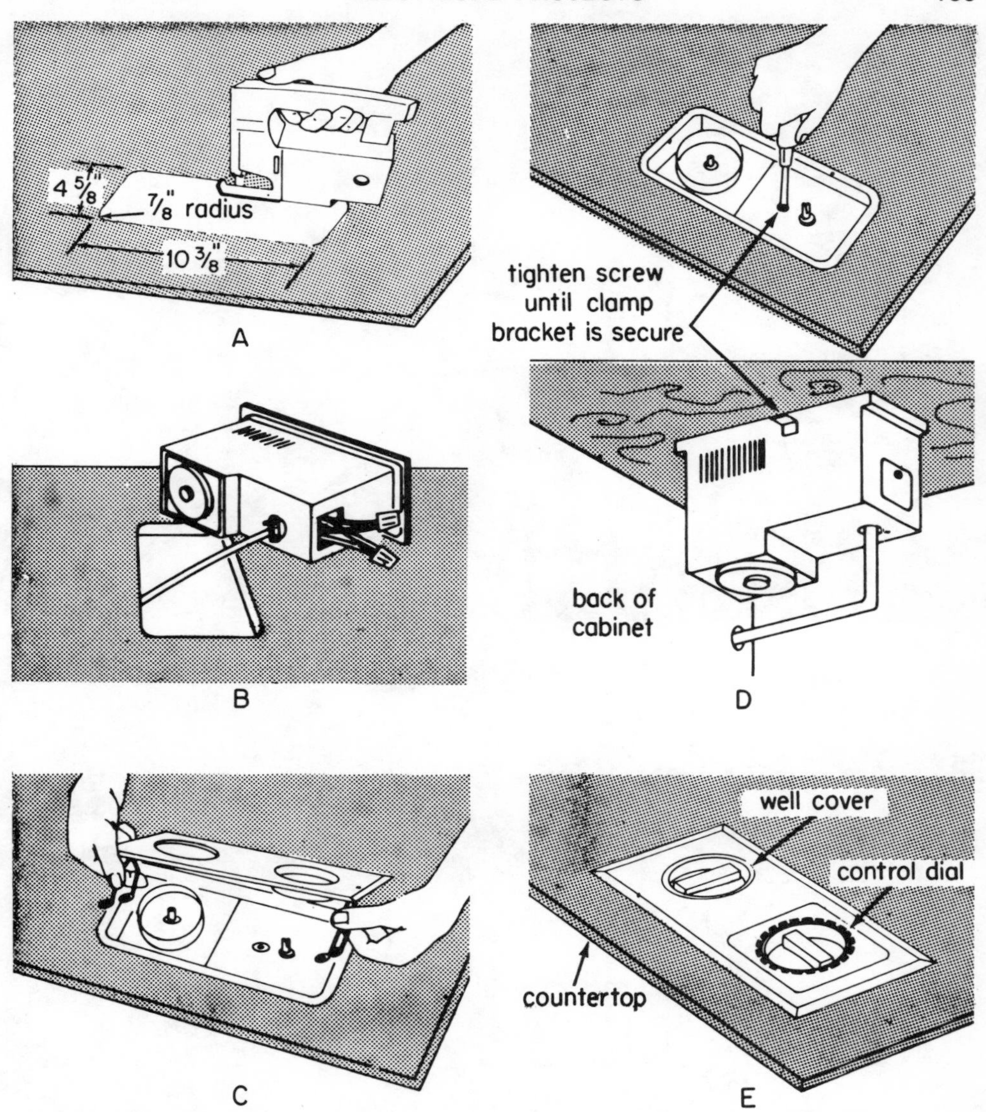

Fig. 13-13. (A) Cut a hole for the unit 10⅜ inches long and 4⅝ inches across. Corners should be rounded using a ⅞-inch radius. (Manufacturer supplies a template.) (B) Run cable up to unit from below, attaching black wires to black, white to white with wire nuts. (C) Remove the top plate from the unit by squeezing the tension springs. (D) Turn the mounting screw clockwise to secure the unit to the counter, until bracket is snug, but not too tight. (E) Replace the plate cover and set control dial in place. Use the motor well cover when not in use.

on the old ceiling, or to the gridwork of the new ceiling (Fig. 13-15). In either case, there must be at least 5 to 6 inches of clearance between the old ceiling and the new. If attaching to the old ceiling, it's economical to use 8-foot fluorescents over large areas of all-luminous panels.

When planning a kitchen area, see Chapter 11 for recommendations as to footcandles and position your lights accordingly. In general, 150 footcan-

Fig. 13-14. Although kitchens are the primary site for luminous ceilings, they can also be attractive and useful adjuncts to a basement rec room (Armstrong), bathroom (Tile Council of America) or sewing room (American Plywood Association).

dles of light are recommended for food preparation. Eating areas require only 50 footcandles. Larger kitchens will benefit from separate ganged switches for the various areas.

Some manufacturers provide matching fluorescent fixtures that fit right into standard 2x4-foot grid openings, and can be purchased from the same dealers (Fig. 13-16). Extra hanger wires are required to support the fixture, as shown in Figure 13-17. Some fixtures simply snap into the grid frame, others are screwed into it. Consult the manufacturer's instructions for complete installation details.

Fig. 13-15. Luminous ceilings are usually part of a suspended ceiling. (Top) Installation begins by installing the edge molding seen in the background. Then the main runners are hung with wires fastened to the joists at 4-foot intervals. (Bottom) Cross-tees are inserted into the main runners. (Top, p. 173) Fluorescent lights are hung on the cross-tees in this system. Translucent panels are installed below the lights, standard panels in the other openings. (Courtesy of Armstrong)

Fig. 13-15 (continued)

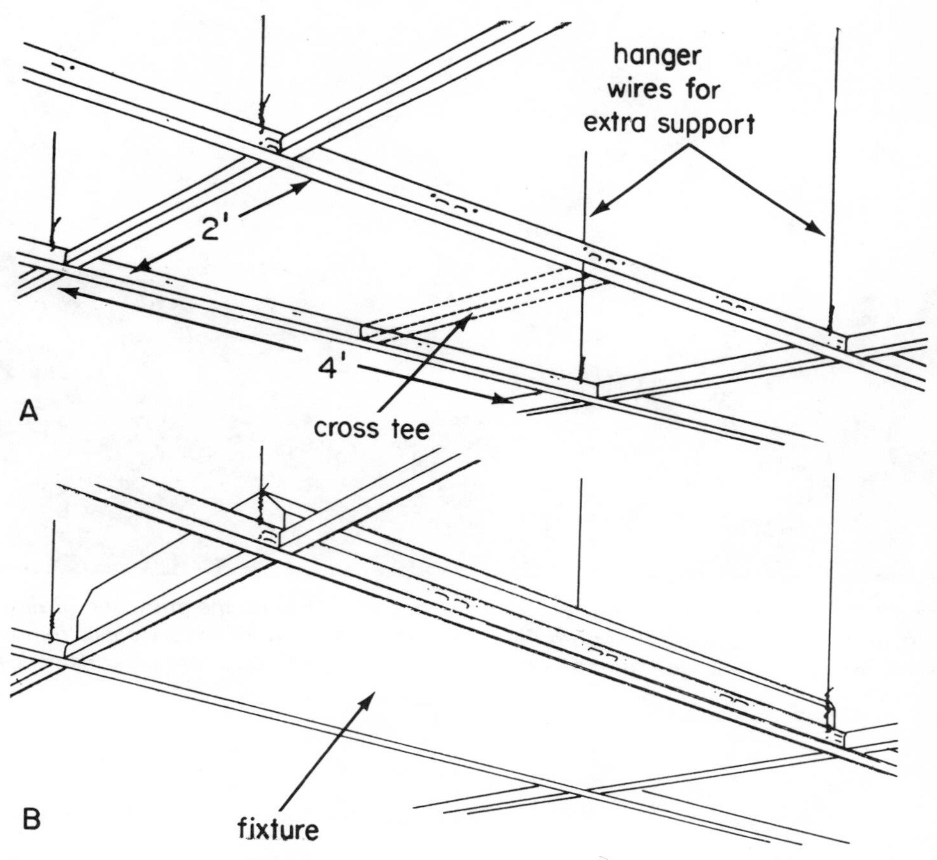

Fig. 13-16. This type of fluorescent fixtures comes as a single unit which rests directly on a 2x4-foot opening in the gridwork. (A) Extra hanger wires are needed to support the fixture. If necessary, remove the cross-tees to enlarge the 2x2 opening to 2x4. (B) Lift fixture through 2x4 grid opening and position to rest in grid framework.

Fig. 13-17. (Above) Some fluorescent fixtures snap directly onto the gridwork. (Below) The luminous panel is then set onto the grid. Both panel and fixture are easily removable and can be changed to another location on the grid. (Courtesy of Armstrong)

Appliance Troubleshooting

Have you ever jabbed and poked and poured abuse upon your toaster, only to find that it wasn't plugged in? Or have you taken your electric can opener to the local repairman, only to have him hand you an inflated bill just for loosening or tightening an adjusted screw? Or for giving you a short course on how it's supposed to be used?

When something doesn't work correctly, don't panic; try the on-off switch! Check the basic things first—like the plug. Some of them are obvious—that's why you forget them.

This chapter is in no way intended to turn you into an instant appliance repairman. People go to school for that, or at least thoroughly research the many books available. What can be done here, perhaps, is to save an unnecessary—and expensive—repairman's call. If you use the troubleshooting charts given in this chapter you can at least determine whether you really need a service call or not. In many cases, the difficulty may be easily remedied; in many others, it may not. Sometimes you can do it yourself, often you cannot. But you should gain "wisdom to know the difference."

There are hundreds of different types of appliances on the market, all made by many manufacturers. While the wiring and operation of these vary greatly (and you are best off if you have a manufacturer's manual for each appliance), there are some basic things in common. Electrically, they can be extremely simple. Heat-producing appliances, such as toasters and waffle irons, are nothing more than heating elements connected to electrical power. Switches and thermostats complicate them a little, but it is easier to troubleshoot and repair them than it is to repair most large motor-driven appliances, such as clothes washers and dishwashers.

Motor-driven appliances are also electrically simple, but can be mechanically complicated. Here again, a manufacturer's manual is needed to understand the operational idiosyncracies of the individual appliance. Some appliances (a rotisserie or heater, for example) combine heat-producing and motorized functions.

As in any repair work, the first thing to do is check the obvious. Avoid panic and jumping to conclusions. Don't assume, for example, that you need a new toaster, when the real problem is that it wasn't plugged in. Don't replace the motor when new brushes are all that is really needed.

A General Caution

Before operating any appliance, make sure that you read the instructions carefully. Save that package of instructions, repair parts, warranty information, and other materials that come with the appliance. They may prove invaluable later.

When something goes wrong with an appliance, dig out the information package. With usage, we tend to get more cavalier in our handling of appliances. In the beginning, we usually pay attention to such things as overloading a clothes washer. As the years slip by, an extra pound may slip in here and there, until we are someday severely overtaxing the capacity. Reread the instructions. This may solve the problem.

In some cases, we may be trying to get the appliance to do something it wasn't intended to do. Has someone used the appliance who wasn't familiar with it—a child, a visitor, the cleaning woman? New users should be thoroughly briefed on how to operate all home appliances.

Before You Pull It Apart

Workman, spare that screwdriver! Before plunging in to take something apart, stop, look and listen. Consider all the possibilities.

Check Sources of Difficulty

External Power Failure

Is the plug in? Is there power to the cord? Both are easily checked, the latter by using a tester at the receptacle. If that doesn't work, there may be trouble either at the receptacle or in the line itself. If you know what else is on the circuit (as you should), see if there is current there. If you don't know, check the fuses or breakers to see if any of them has blown.

When replacing the fuse or tripping the breaker restores power to the line, but turning on the appliance causes it to retrip, the problem could be an overload. Nonheat-producing small appliances should not ordinarily, unless there are other appliances on the same circuit, cause overload. If you are uncertain whether the problem is overload, try the appliance on a different circuit that you know can handle it. If the fuse blows again, then you can be reasonably sure that the appliance has a short. Also check for low and high line voltage, but note that this spells trouble for many electrical devices. The power company is required to correct such situations, even if it means changing a pole transformer or stringing new lines to guarantee proper levels at the entrance panel.

Physical Damage

If an appliance has been dropped recently, or was in a fire or flood, any damage is probably obvious. If fire (other than an internal short) was the culprit, there should be coverage under your insurance policy. In that case, your problems are over. The insurance company should pay to repair or—more probably—replace it.

Flooding is a different matter, flood damage rarely being covered by insurance, except under special government-backed policies. If an appliance was damaged in a flood, you should most probably discard it. You can possibly save it by cleaning it thoroughly and baking the electrical parts in a low oven (180° F). Check all the contacts for cleanliness and replace any damaged insulation. Tighten all connections, lubricate moving parts and remove all traces of rust with emery cloth or rust-removing fluids. It's worth a try, but chances of successful rehabilitation are not good.

An appliance that has been dropped should have connections checked. See that everything lines up properly. If a plastic cover has been broken or smashed, look for tiny particles in moving parts. Dents in metal surfaces may be pressing against something inside, causing a short or moving parts to jam. Parts may have to be replaced, but chances of recovery are good.

Use Your Senses

If none of the above seems to be the problem, it is time to use your senses—and your head. If you smell something burning, insulation may be melting. The appliance should be disconnected immediately and checked. Remove the cover and check for burned-off or smudged insulation on the wires. If anything looks questionable, replace it just to be safe.

Noise is another giveaway. A rattling noise in a motor-driven appliance may just necessitate tightening a few screws. A grinding noise on gear-driven machines indicates worn gears or bearings. Squeaking is a sign of neglected oiling. Loose or slipping drive belts have characteristic flopping or humming sounds.

It goes without saying that the eyes are also an all-important tool for locating problems. Check for loose connections, frayed insulation, scorched parts, and so on. Check the indicator lights, if any, to see if they point to the problem. But first make sure the indicator light itself hasn't burned out.

Do-It-Yourself or Professional Repair?

Once you have located at least the general source of difficulty, the next question is whether you should tackle it yourself or call in the local appliance service. In large part, of course, that question is unanswerable here. It all depends on how skilled you are at diagnosis and corrective surgery.

Check Warranty

Before you even consider this option, however, check to see if the appliance is still under warranty and what is covered. If your luck is typical, the warranty will have run out the day before the trouble occurred. But you may be more fortunate than most. All, or most, of the repair cost may be covered by the manufacturer.

Even if the appliance is still under warranty, you may find that it is easier and less expensive to fix it yourself. In spite of reforms over the past several years, warranties are more generous in what isn't covered than what is. A brand-new machine may often be exchanged quickly at reputable stores, but if you've used it awhile, the warranty may require that you pack up the appliance and send it across the country, accompanied by a rather stiff "handling charge." You may find that parts are warranted while labor isn't. The part cost is often minimal while labor charges usually run high.

The best course is to inspect the warranty carefully when you *buy* the appliance, and avoid those that involve long and costly by-mail returns. If the warranty is intact and there is a factory-authorized repair shop in your vicinity, there is no point in trying to fix the appliance yourself.

Taking It Apart

If you've saved the instructions and wiring diagram from the manufacturer, you're way ahead. Exploded views and other drawings will show you how to take apart the appliance and, more important, put it back together. Taking things apart is usually the easy part. If you don't have the manufacturer's instructions, make your own diagram as you go along. It will be a lot simpler when it comes to the hard part: putting it all back together correctly. Always set the pieces on a newspaper or cloth in the order in which you take them apart.

Locate Causes of Failure

Once you get the appliance housing apart, the electrical system should be evident. The first thing to check for is loose or broken wiring. On heat-producing appliances, inspect the heating element. Small appliances generally have nichrome (an alloy of nickel and chromium) coils or ribbons. These can break and are usually replaceable. If the break is near the end of the element, it is sometimes possible to stretch the element a little and reattach the unbroken section to the screw. But don't do this unless the removed section is very small. A shortened heating element will give off more, not less, heat and could be dangerous.

Use Testers

A great helpmate in diagnosing appliance ills (among other electrical problems) is a neon or continuity tester, as described on pages 53–54. Even more valuable, but more difficult to use, is a volt-ohm-milliam-meter (VOM). Since the VOM is a highly specialized tool, I won't get into its usage in this book, but remember two things:

- If the VOM doesn't come equipped with alligator clips, get a set of test leads, or make a pair by wiring the clips to a couple of 2-foot lengths of lamp cord.
- It is often necessary to use a VOM "under load," so keep both hands on the test prods, or keep one hand in your pocket when testing. If you position both hands across a live circuit, your body itself will wind up as the tester—a condition uniquely hazardous to life and limb.

Appliances You Should Not Repair

Certain appliances come into close contact with our bodies, for example, electric blankets, heating pads, and electric toothbrushes. If you have a problem with items of this type, either return them to the manufacturer (whether under warranty or not) for repair, or get rid of them. A repair mistake can be deadly, so the most you should do is replace a damaged cord or plug. An exception is an electric shaver, where the blades are frequently worn or broken. The blades are quite easily replaced.

Simple Heat-Producing Appliances

The first group of appliances discussed in this chapter are relatively easy to troubleshoot and repair. The symptoms are usually readily apparent, the cause pretty evident, and the cure generally an easy one.

The reason that these appliances are so easy to fix is that they are basically a heating element of one sort or another (a resistor) in the form of a coil or plate made of nichrome or other special alloys, and a cord to provide current to the heating element.

Practically all heat-producing appliances are governed by a thermostat, although some older models may not be so equipped. Thermostats on an appliance work on the same principles as those used for furnaces, as explained on page 104. The only real differences between a pop-up toaster, a waffle iron, and a steam iron are that they are mechanically adapted for specific purposes.

Sometimes, heat-producing appliances are combined with motors. Some examples are a rotisserie and a clothes dryer. These are discussed later in this chapter.

Dry and Steam Irons

A standard dry iron is really nothing more than a resistor, or heating element, drawing current from the circuit (Fig. 14-1). The amount of current is determined by a thermostat, which is controlled by the dial on the top. Most irons are marked linen, wool, permanent press and so on, but these are really translations of the degree of heat being put out by the heating element.

When disassembling an iron, mark where the indicators point. This is important. Then, when you put the iron back together, the dial will line up correctly with the thermostat settings.

On both dry and steam irons, a frequent source of trouble is a frayed cord, particularly at the plug and at the cord sleeve next to the iron. (See Chapter 5 for cord and/or plug replacement.) Breaks elsewhere in the cord can be suspected where the exterior looks worn, burned, or is soft and bends too easily.

Another frequent problem is melted synthetic fabrics or cooked starch, which leave a sticky film on the iron bottom plate. Clean the plate with a damp cloth and wipe it dry. If this doesn't work, rub gently with fine steel wool or a piece of very fine emery cloth. Slight scratches will not harm the plate, but burrs will damage clothing and should be removed with emery cloth.

Steam irons are especially plagued by hard water, which is found in most parts of the country. The minerals in hard water accumulate and clog the small steam portholes and passages. The problem can be prevented by using a water softener or distilled water. Rain water, melted snow, or defrosted water and accumulated ice from the refrigerator are good sources of mineral-free water. If it is too late for any of these preventive measures, fill the tank with vinegar, which will dissolve the mineral deposits. It may be necessary to heat the iron with the vinegar inside, and to repeat the process until the iron is clean.

Toasters

Like an iron, a toaster is a very simple mechanism, consisting of a heating element and a cord in a housing, along

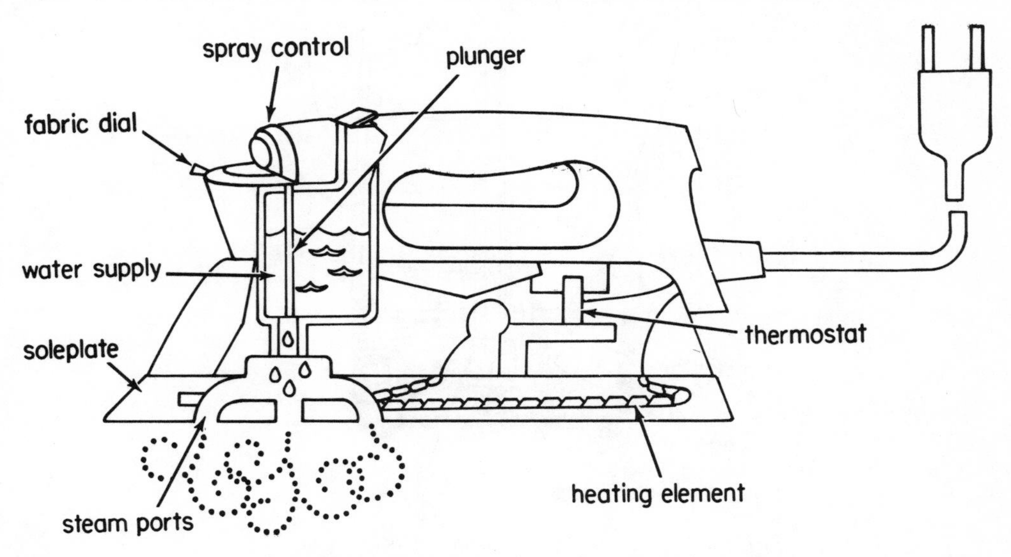

Fig. 14-1. Cross-section of a steam iron.

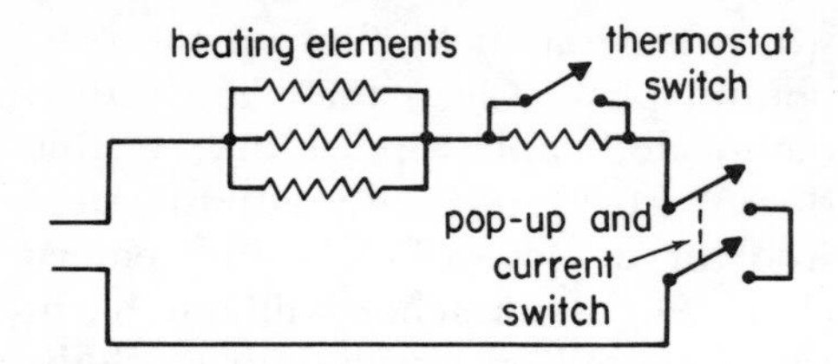

Fig. 14-2. Wiring diagram for a typical pop-up toaster.

with a timer and/or thermostat. A two-slice automatic has a double-sided heating element in the middle and one single element on the outside of each slice. Four-slice and larger models are similarly arranged.

In the schematic diagram (Fig. 14-2), the thermostat is represented by a resistor plus a switch. The other switch represents the combination pop-up mechanism and current shut-off.

Toaster problems are frequently caused by crumbs collecting in the mechanism. An ounce of preventive maintenance, in the form of a thorough cleanout every few months, is strongly advised. Most toasters have a removable plate on the bottom for crumb-cleanout. There is not much to repair in an automatic toaster. Occasionally a heating element is ruined. At times a new thermostat, timer, or cord will be needed. It is easier and cheaper to buy new parts than to repair the old.

Coffee Makers

Coffee-making and coffee-drinking are sacred rituals in some homes and keeping the coffee pot in top shape is of vital social importance (Fig. 14-3). But the most expensive or efficiently operating pot won't guarantee good

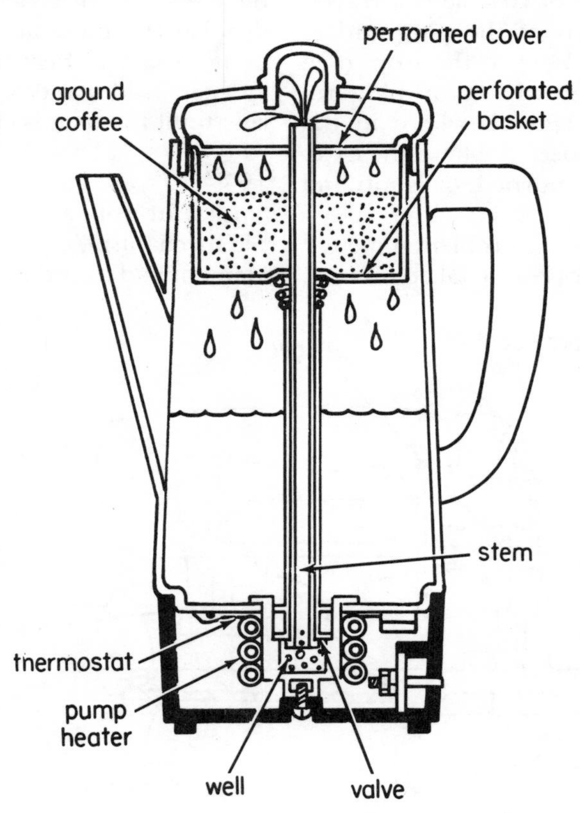

Fig. 14-3. Anatomy of a typical coffeemaker, percolator style.

coffee. The problems that plague coffee *klatchers* are usually due to two causes: defective water heating or built-up flavor-killing deposits.

Defective water heating results in coffee that is too weak (flavor left in the beans) or too hot (bitter oils extracted along with the flavor). Water should be between 175 and 190° F. when it passes over the coffee. If it isn't, check the diagnostic chart.

Deposits can be built up in the coffee pot either from hard water or improper cleaning, or both. Hard water will leave mineral deposits and improper cleaning will leave coffee solids containing bitter oils. A regular schedule of "boiling out" the pot with baking soda will do much for the flavor. A strong vinegar solution will dissolve lime, the most frequent mineral deposit.

The perking action of a coffee pot is not caused by boiling water, as is commonly believed, but by steam. Steam is generated under the basket stem base in the pump heater well. The steam "pumps" water up through the percolator tube in spurts. The water then splashes down over the coffee grounds, extracting their flavor. The flow is controlled by the pump check valve, a loose disc at the bottom of the stem. As water is perked up, more water flows into the well valve, where in turn it is pumped up to the top of the tube and out. It's important on all pump-type coffee percolators to keep the disc and valve seat clean, smooth and unscratched.

Actually, most coffee pots have two heaters. One starts the perking action, while the other acts more slowly to heat up the water. When the water is heated, the thermostat keeps the coffee heated as long as the plug is in. Up to that time, the thermostat remains closed so that full line voltage can reach the pump heater.

The newer coffee makers on the market operate in a similar fashion, except that the glass "pot" sits on a heating plate. The water is heated in a separate base reservoir, from which it is delivered to the pot.

Frying Pans

The electric frying pan is another of those basically simple appliances which contain no moving parts—it's just a heating element and a cord (Fig. 14-4). There is one important distinction among otherwise similar brands that should be printed clearly on the name plate—immersibility. A new user may have to check that impulsive habit of dunking the frying pan in water. Some of them can be dunked, but check the label first.

If you submerge the nonimmersible type, take off the bottom plate as soon as possible and dry off the unit with a fan or vacuum cleaner blower. The best tool for this job is a blower-type hair dryer.

Electric pans do not go wrong easily. Second to unsanctioned immersion, the usual causes of malfunction have to do with a similar failure of the user to realize that an electric frying pan is

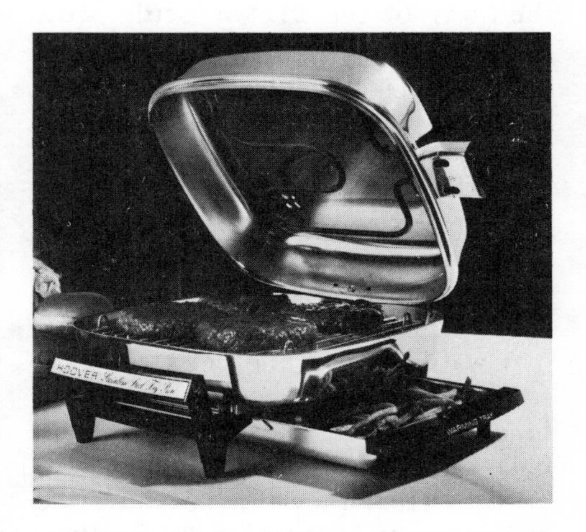

Fig. 14-4. Most electric fry pans consist of a heating element and a cord, plus other features such as the warming pan on this model. (Courtesy of Hoover)

more than just a frying pan with a cord on it. Careful reading of instructions should make this clear, but many people simply do not believe in reading instructions.

These pans cook faster at lower temperatures, and the habit-conscious user often turns the heat too high for too long. The result is sticking and burning, and a difficult clean-up job. The best way to clean an electric pan is to heat several cups of water in the pan for a few minutes, then unplug the unit. Scrape out the residue with a wooden spoon or spatula, then wash out the inside with hot water and a mild soap.

Strong detergents, alkaline cleansers, and abrasives should be avoided. Like the old iron skillet, the surface of the frying pan must be seasoned; its porous surface should be sealed with fresh shortening every time such cleaners have been used.

Although it is a sturdy appliance, the electric frying pan is more susceptible to external damage than some others. Frequently, a hard blow will change the thermostat calibration. To check, note the setting at which water boils in the pan. If it is within 20 degrees either way of the 212° F. setting, the calibration is close enough.

When reassembling an immersible-type frying pan after making any repairs, take care to reseat gaskets properly. If a gasket looks worn or won't seat properly, replace it with a new one. The cost is negligible.

Waffle Irons

A waffle iron is another electrically simple appliance—a resistor connected to a plug (Fig. 14-5). Many models have a thermostat and removable grids, so that they can be converted into a smooth-surfaced grill. The thermostat tells you when the grids are hot enough to use, and keeps the inside at a constant temperature.

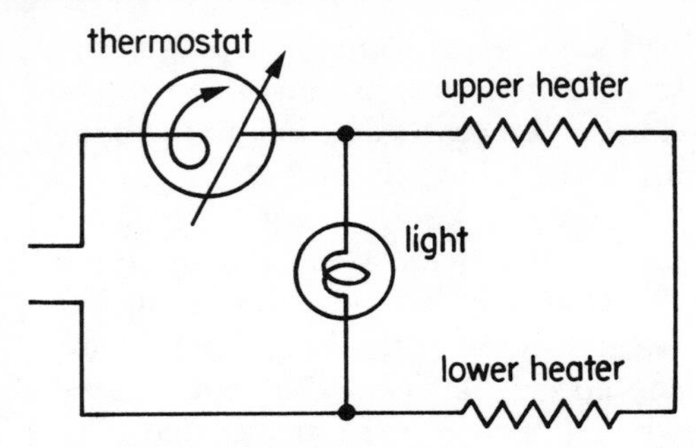

Fig. 14-5. Wiring diagram for a waffle iron shows that it is not much different from other heat-producing appliances. The principal difference is two heating elements instead of one.

The entire electrical system is readily accessible on most models. Typically, the grids lift right out and there it is: a long heat-producing coil on top and bottom, a light, and a thermostat. When the iron is hot enough, the thermostat breaks the circuit, shutting off the light.

Visual inspection should reveal any breaks in the coil. The coil should be replaced unless the break is within an inch of the end post; you can usually stretch it that far and reconnect without any damage. Be sure to tighten screws all the way, and never allow looseness in the coil—direct contact of the coil with the plates could be fatal. After any such reassembly, the unit should be checked for a short. Put one lead of a tester on the power terminal and the other on the shell to see if any power has escaped. If the light lights, look out! The old, usually round, waffle irons (such as you might have inherited as a family heirloom) are wired in parallel, so it is easy to determine if the coil is defective. One half will work and the other won't.

The waffle grids can be black or shiny or almost anything except unoiled. If a grid has been burned too badly, it requires patience and hard work to remove all the batter. But it must be removed. Don't use a detergent unless absolutely necessary. If you must use a detergent, the oil will be drawn out of the pores, and the iron must then be seasoned again, just like a new one. As an extra precaution, add a bit more cooking oil to the batter.

Ranges and Ovens

Electric ranges and ovens are super hot plates, basically no different from smaller heat-producing appliances (Fig. 14-6). Complications arise because of timing and self-cleaning devices, but ranges, whether divided into ovens, broilers, and counter-top cooking elements, or combined into one unit, should be treated as other similar appliances.

You can replace the heating elements in each unit, fix loose connections, and so on, without fear of complications, but be very sure that the

Fig. 14-6. Many modern homes have a separate cooktop, with individual ovens built into the wall. Both units shown here feature a glass-top "counter that cooks." (Courtesy of Corning Glass Works)

current is off. Remember that you're dealing with 240 volts. If you work on the wiring, make sure that you reconnect the wires correctly.

Ordinarily, however, an electric range requires only periodic cleaning to keep it in good shape. And keep the manufacturer's manual around in case anything goes wrong. Even if you call in a professional, the manual can help him track down the trouble more quickly, thus saving you some of his expensive time.

Electric Motors

This seems to be as good a place as any to introduce the subject of electric motors. In most cases, a burnt-out motor should simply be replaced with one of the same type. For small appliances, such as an electric shaver, it is easier and less costly to get a new one. More costly appliances can often be saved with a new motor. In some cases, new brushes will do the trick. Motor repair, however, is a topic all its own, and I have neither the space (nor the knowledge) to delve into that here.

You should know, however, that there are two basic motors, synchronous and brush-type. It is well to have at least a passing knowledge of each if you intend to do any appliance repair involving motors.

Synchronous Motors

Synchronous motors are small and simple (Fig. 14-7). They consist basically of two elements, a field and a rotor.

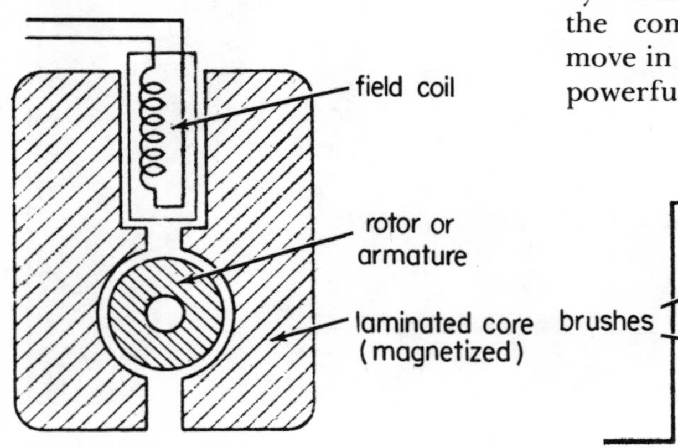

Fig. 14-7. Schematic diagram of a synchronous motor.

The field is a U-shaped laminated frame around which are wrapped many turns of fine wire. The wire is plugged into a power source, creating a magnetic field. In an AC circuit, this field changes direction 120 times a second, causing the rotor to revolve. This motor is excellent where a constant (synchronous) speed is desired, such as a clock.

Brush-type Motors

Brush-type motors are also called "series-wound universal motors"—universal because of their wide possibility of application. In this type, two fields are employed. Instead of the simple rotor found in the synchronous motor, there is a more complicated version of the same thing called an armature. The armature moves opposite to the field by means of a commutator, which gets its drive from the coils inside each of its bars. Impetus is given by "brushes" which cause one bar of the commutator after another, to move in opposition to the field. A very powerful push is thereby created.

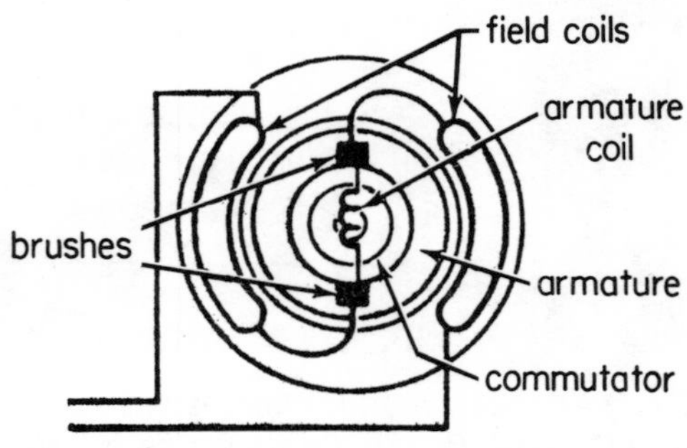

Fig. 14-8. A typical brush-type motor.

The brushes are made of soft carbon and frequently must be replaced. Be sure to use exactly the same size brushes as replacements. It is also important to know that each bar of the commutator has its own coil and is connected to the next. If the commutator bars operate unevenly, it is usually advisable, and not too costly, to replace the entire armature. Occasionally, sandpapering the copper bars of the armature will temporarily restore efficiency.

Combination Heat-producing and Motorized Appliances

This group of appliances would logically fall in the first group of simple heat-producing units. The only difference is that each has a motor to perform a service unique to that appliance. The main purpose of the motor in this type of appliance is to blow the heat in a certain direction, or to turn a part, such as the spit in a rotisserie or a clothes basket in a dryer.

Basically, these are not hard to repair. A clothes dryer is more complicated because of the timer mechanism. It is also sometimes frustratingly difficult to take a dryer apart. The manufacturers often seem to design them so that you can't tell how to get inside. Once you're in there, however, you can replace a heating element or a motor without too great a problem.

Portable Heaters

In these days of frighteningly escalating fuel costs, a portable electric heater can be much more economical than it was formerly. Electric heat is becoming relatively less expensive than it used to be. In fact, it is a wise idea to keep the central heat down, and use portable electric heat in rooms only where the family congregates, such as the family room or basement play room.

Once again, we have another appliance that is simply a task-specific heating element with a cord. Most units also include a motor-driven fan which circulates and directs the warm air from the heating element. Fan repairs are similar to those for a standard electric fan. A common defect in fans is excessive noise, which is usually caused by bent blades and can be cured by putting them back in their original configuration.

It is recommended, incidentally, when purchasing a portable heater, that you buy one with an automatic switch that shuts off the fan if it is tipped over. Dangerous burns or fires can be started by heaters which are knocked down by children or animals. The switch cuts down on this danger.

If the heater does not shut on and off when you think it should, the thermostat may be at fault. To adjust, remove the knob and adjust the set screw in the center of the shaft. If adjustment doesn't do the job, a new thermostat may be in order. Most heaters have a removable back that exposes the wiring for point-to-point checkouts. While you're taking the heater apart, clean out any dust with a vacuum cleaner and a small paintbrush. It is also important to clean the reflector regularly. In addition to dry wiping, it should be cleaned periodically with a mild soap and wet sponge to increase its efficiency.

The heating element can be visually inspected by removing the grille and taking a look. A large break in the element should be easily observed. Do not shorten the resistance wire and connect to the nearest terminal if you don't want to blow a fuse. Replace the element.

The wiring in an electric heater is almost always asbestos-insulated. Because of the recent findings on asbestos as a cancer-causing agent, use glass tape if you have to reinsulate. Any other material is apt to soften, smoke, and smell as it gets hot. (See below for additional warnings on asbestos.)

Hair Dryers

The hair dryer has two basic parts: a motor-driven fan and a heating element. The fan blows air over the element, which gets very hot, and the warmed air flows out the nozzle. Up to this point, all dryers are pretty much the same. From the nozzle onward, they divide into two types, the hand-held and the hood. Although they look radically different, their operation is almost identical. The difference is that on the hood models, a length of plastic hose is attached to the nozzle, along with a hood that has a number of small openings to diffuse the warm air. Improved hand-held models seem to be replacing the hood types.

Since the heating element gets quite hot, a hair dryer has a dual switch that makes it possible to turn the motor on without the heater, but not vice versa.

If neither motor nor heater is working, it is a good bet that the trouble is in the cord assembly at a common connection to both motor and heating element. If either one works and the other doesn't, start investigating the nonworking area and let the working one alone. When you have located the source of trouble, replacement of the malfunctioning unit is usually indicated, unless it is simply a loose connection. On hood types, you occasionally find a leaky hose. Plastic tape does an effective, temporary patching job.

The Federal Trade Commission (FTC) has determined that the asbestos insulation in many hair dryers can be hazardous to your health. Asbestos is one of the most deadly of the carcinogens (cancer-causing agents), and the blower unit can relay some of the fibers into your respiratory system. Manufacturers of the offending types will repair or replace these dangerous models, if you still have one.

For the same reason, we do not recommend do-it-yourself repairs of this type of hair dryer. By all means, return it to the manufacturer. You may get a free dryer in the process, even though yours is old and defective. Don't fool around with asbestos, even if you're skeptical of all the publicity about cancer-causing agents. This is no laboratory finding based on saccharin-ingested mice. Asbestos workers have an incidence of lung cancer at least 16 times that of the ordinary person.

Since the only repair or replacement options for hair dryers are already given above, there is no diagnostic chart for this appliance.

Rotisserie Oven/Broilers

A typical rotisserie consists of a thermostatically controlled heater element, a motor to operate the spit, and a timer to stop the cooking after a preset period (Fig. 14-9). Spills from overcooking are a common cause of trou-

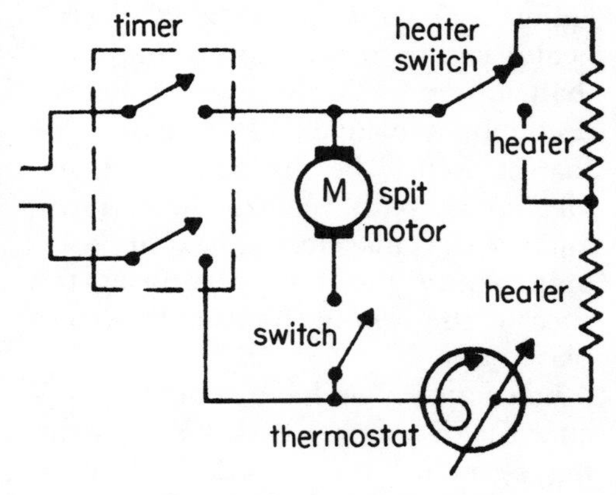

Fig. 14-9. Basically, a rotiserrie is a couple of heating elements with a motor-powered spit.

ble with an electric rotisserie. If this happens, it is much easier to clean it up right away than to remove baked-on deposits.

Make sure the oven is cool and unplugged before washing. The glass, particularly, may break when wiped or washed while still warm. Dry by hand, then turn on the oven to 250° F. and let it run for awhile. This should prevent rusting.

Clothes Dryers

Electric clothes dryers use both 120 and 240 volts. The lower voltage controls the motor, the higher is for the heating element, which consumes about 5,000 watts. The heater is turned on by a centrifugal switch, which means that the motor must be started first and that the drum should be spinning quite rapidly to provide enough force to open the starting switch and close the heating switch at the same time. This is done so that the heating element does not operate with the clothes in a stationary position, causing them to burn.

A gas dryer works similarly, except that gas instead of 240-volt electricity provides the heat. Most dryers have an electric solenoid starter to ignite the gas (some gas models have pilot lights). An electric motor also powers the tumbling action.

Sometimes the motor drive belt or heating element gives out. Contacts may need cleaning or replacing. Occasionally, the door switch fails. Careful inspection and disassembly should reveal the problem. Whatever you do, don't be careless with the wiring in an electric dryer. The voltage here is at a lethal level. Need I say it again? Make sure the current is off before fooling around.

Dishwashers

A dishwasher works by spraying hot water and detergent with enough force to melt, emulsify, and dislodge various scraps of food left on the dishes. Ordinary tap water is often not hot enough, so in some models the drying element further warms both the water and the inside of the machine. A timer controls such various aspects as the drying element, water valves, detergent cups, agitator, motors, and pumps. There are several cycles, and some timers are extremely complex if they have extras such as rinse and hold or dishwarmer. (As noted in Chapter 12, these are very energy-wasteful cycles, and should be avoided when buying a dishwasher.) The more complex the timer, the more advisable it is to call in the serviceman when something goes wrong.

The most common complaint about a dishwasher is that it does not get the dishes really clean. This is more often the fault of the operator than the machine. Dishes must be properly prepared before they are put into the machine. The best way to do this is to clean the dishes of grease and loose particles and put them into the washer immediately after use. Such residues as cooked eggs and tomato sauces are very difficult to remove if left standing, so dishes should either be scraped or rinsed at the sink, or rinsed in the "rinse and hold" cycle if there aren't enough dishes for a full load, preferably the former.

Improper loading is another common error. Follow the manufacturer's directions closely, and always be sure that the soiled side of the dish is facing the spray. Dishes that are jammed in with no space between them leave no room for the water to do its work, and a large dish in front of the detergent cup will interfere with proper operation.

There are two general types of dishwasher agitators. The impeller type stays below the water level, propelling the water over and around the dishes.

In the spray-arm type, the water level stays below the arm, and the pump drives the water up through the arm and through the holes onto the dishes. In both types, as the water is forced throughout the unit it mixes with detergent from the cup and removes dirt and grease. After several washes and rinses, the water is pumped out and the heating element dries the dishes.

Common and fixable operational problems with dishwashers include food particles in the pipes and the pump, worn solenoids and gaskets, defective door switches and heating elements, and broken detergent cups.

Motor-Driven Appliances

The number of motor-driven appliances is legion, and more are being developed all the time. If you think we've seen the end of them, consider what your grandparents would have said if a futuristic sage had predicted electric shavers, toothbrushes, carving knives, and can openers.

The smaller appliances, such as those mentioned above, are usually driven by a small motor, usually synchronous, and have common problems. In many cases, they are dropped, blades wear out, cords become worn. Replacement of the defective part or the entire unit is usually advised. A little snooping around inside the unit, or a visit to the local authorized repair dealer, is the best solution to problems with this type of appliance. If that won't do the trick, it is usually not worthwhile to repair the appliance. It is often more economical to buy a new one, so I won't get into involved diagnostic procedures for less expensive appliances.

Mixers

When the beaters of an electric mixer fail to turn, and if a hum indicates that the motor is running, it usually indicates bent or defective beaters or worn beater gears. Visual inspection should tell you whether the beaters themselves are okay. If not, they are easily replaced. Replacing gears is a little more complicated. You will have to take the mechanism apart so that the gears may be inspected. You will probably find that the gears are stripped, but it may be just a case of plain wear and tear.

When replacing the gears, it is important to position them so that the beaters do not strike each other. Test the machine with the beaters in place before putting it back together. If they touch, adjust until they are completely independent of each other.

Another item that often needs to be replaced (in pairs) is the motor brushes. Most mixers have brush caps on the outside of the motor housing so that inspection is easy. If they are worn to less than ⅛ inch from the spring, they should be replaced. Check the springs, too, to see if they are damaged or weak.

If the brushes are not smooth, clean and curved to the surface of the commutator, you may find that the armature is defective. Before checking that, make sure that both brushes have been removed to avoid breaking them. If the trouble is in the armature, it is best to get a new one, or a new mixer.

Blenders

After dropping and breaking, the most common blender problem is damage to the cutting blades. These will chip or curl back if nonblendable objects are fed into it. (A common misuse of a blender is to mix drinks with whole ice cubes. A blender is not an ice crusher. To use ice in most blenders, you have to crush the cubes first.) Blades are changed by holding the knife shaft with a wrench and turning the spiral nut on the blades clockwise, then removing, replacing, and retightening.

Leakage in a blender is caused by a poor seal between the jar and the bushing assembly. Sometimes the bushing can be tightened, but usually it is best to replace the gasket—a simple matter. The gasket is easily replaced by unscrewing the base from the container, and then taking off the nut and bushing. Install the new gasket and reassemble.

Vacuum Cleaners

Vacuum cleaners are divided into two distinct types, upright and tank. Their differences are based more on styling and customer preference than on any basic mechanical dissimilarity. Both types consist of a motor-driven fan that creates a suction, thereby drawing dirt from carpeting and other materials. The larger upright models have a brush that "beats" the rugs and loosens dirt. This helps them to do a better job on rugs, but the tank-type is generally conceded to have more versatility and drawing power.

The fan section of the housing is sealed off from the motor area so that none of the dirt gets into the motor. The dirt is sucked into a cloth or paper bag (usually disposable) through an aperture. The bag is of such a consistency that it allows the air to escape, yet retains the dirt.

One common reason for poor and noisy operation of a vacuum cleaner is brush-bearing wear. The beater brush rotates at very high speed and dirt is literally forced into the bearings, caus-

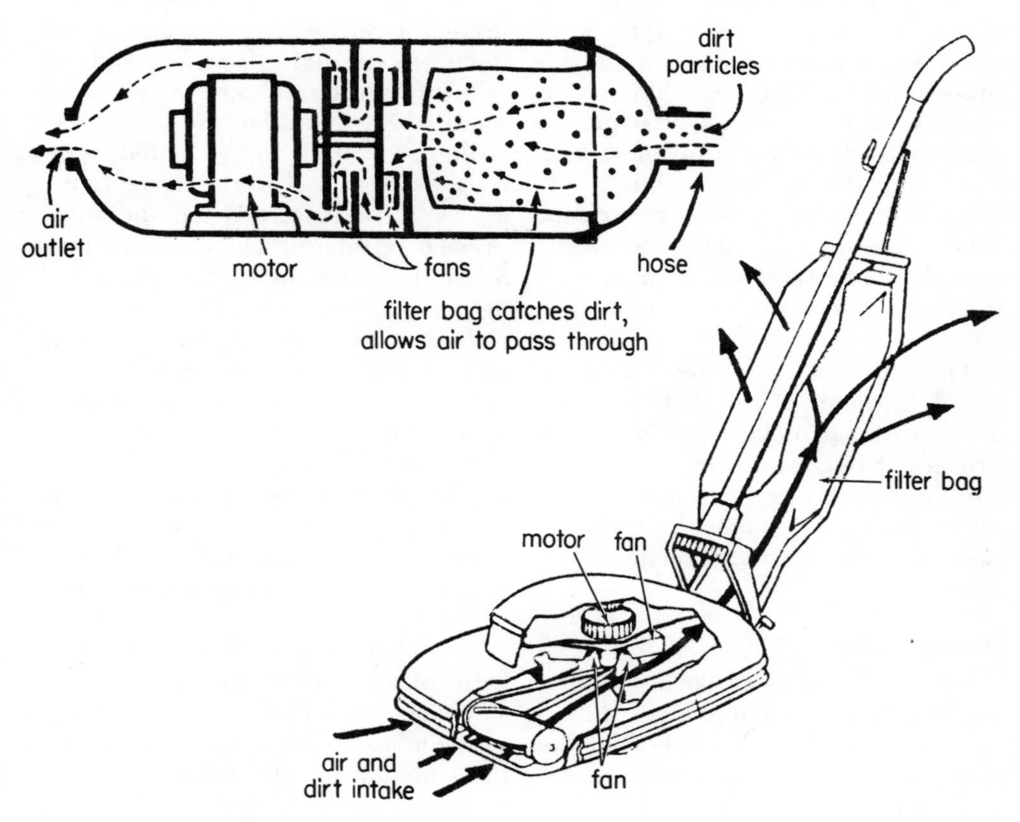

Fig. 14-10. Anatomy of the two basic types of vacuum cleaners.

ing them to lose their lubrication. Frequent cleaning will help, but bearings will need periodic replacement. Very often, when a lot of foreign objects such as string or paper clips are picked up, the fan becomes jammed and the machine will lose suction or fail to run at all. This happens often in the smaller models. The housing must be opened up (usually not difficult) and the offending substance removed. Occasionally, a foreign object will damage a fan blade, causing it to run very noisily. Replacement is indicated.

Polisher-Scrubbers

These machines clean, polish, scrub, and shampoo. They work on hardwood or rugs, and require very little effort on the user's part. Most polishers can also be fitted with attachments to adapt them for floor sanding.

The polisher-scrubber consists of a high-speed motor and a spur gear train. At the end of the gearing-down process are two brushes. Maintenance consists of keeping the attachments reasonably clean. Build-up of dirt or wax can affect the machine's efficiency, so brushes must be kept clean. When polishing brushes are worn, they should be replaced in pairs, even if one seems to have some working life left.

The universal motor requires periodic lubrication of the shaft bearings and occasional replacement of motor brushes. Noise when the machine is running usually signifies some motor malfunction. Other warning signs are poor performance and excessive vibration.

Refrigerators

When you boil water, you are changing H_2O from a liquid to a gas. To accomplish this, heat is applied to the water. To change water into ice, heat is removed, the reverse process. To test this, rub a little alcohol on your hand and notice how cool it feels. As the alcohol evaporates, or changes from a liquid to a gas, it draws heat from your body. Gases cool when their pressure is allowed to drop.

A refrigerator works on the same principle. (So does a freezer, only at lower temperatures.) The refrigerant, usually Freon, is a liquid that boils at a much lower temperature than water. If you were to put a jar of refrigerant into a box, it would begin to boil, drawing heat out of the box.

In a refrigerator, the Freon is contained in a closed system of pipes, moving through the refrigerator and back again by means of a compressor. The compressor is driven by a small motor. As it goes through the refrigerator, the liquid changes to gas, drawing heat out of the "box." The compressor takes the heated vapors from inside the refrigerator coils and moves them out through a series of coils of smaller diameter. The restricted pipes put the Freon under pressure, thus changing it back into a liquid. The coils, visible in back of the refrigerator, act as a condenser. When the compressed refrigerant changes back into a liquid, it loses heat. The hot air is lighter than the surrounding air, so it rises, thus causing cooler air to move in and remove even more heat. In most refrigerators, this action is accelerated by the blowing of a fan across the coils.

As the refrigerant reenters the box, the diameter of the pipes is enlarged and the pressure is removed. The Freon again changes to gas and begins to remove heat, cooling the box and whatever is inside.

Note that there must be free movement of air around the coils both inside and out. This is usually not a problem inside the box, but very often the coils outside the refrigerator become clogged with dust and dirt. It is important, therefore, to give the coils a periodic vacuuming to keep them in proper working order.

When adverse cooling conditions exist, for example, in hot weather, the door may be opened and closed frequently, thus letting the cold air escape. Set the thermostat higher.

Conversely, when the air in the house is cold, or when there is little inside the refrigerator to be cooled, the cooling action can work too well, and the food may freeze. The cure is to turn the thermostat down.

Except for keeping the coils clean and changing defective parts such as a thermostat, there is not much that you can do for a refrigerator's electrical system. The compressor and motor are usually in one housing, and they are built to last without oiling for at least 10 years. If and when they give out, it is probably time to buy a new refrigerator. Do not attempt any repairs that involve taking apart the coils. If the part cannot be replaced without disturbing the refrigerant, either call a serviceman or replace the whole unit. Refrigerants should not be tampered with or allowed to escape.

Some of the more common mechanical problems are worn gaskets, defective or worn motor mounts, and, occasionally, defective latches. These can be replaced without too much difficulty. Insulation may become worn after many years, but replacement involves tearing out the inside of the box—you will probably do more harm than good. When this is suspected, it is time for a new model.

Automatic Clothes Washers

Automatic clothes washers perform a complex series of operations. First, the soiled clothes are pre-washed, and the water is drained. Fresh water is added, and the clothes are either agitated or spun rapidly and tossed around through the water. This is followed by several rinses, usually with the drum spinning rapidly to get rid of the water. A deep rinse is often added as well. After the final spinning, the clothes are damp dried. They are then hung out to dry or put into an automatic dryer.

Many moving parts and gears are required to carry out the different actions. In addition, most washers have several settings for large and small loads and various water temperatures for different types of fabrics, requiring still more switches plus solenoids to control the hot- and cold-water taps at different parts of the cycle. Further, there should be a safety switch to stop the machine when the cover is opened, and another to shut it off when it is overloaded or unbalanced during the spin cycles.

Obviously, the electrical system of a clothes washer is not to be haphazardly tampered with. The various pumps, gears, and the like make this machine mechanically complicated as well.

Washers take a lot of hard abuse, and their life expectancy is considerably lower than that of a refrigerator, a dishwasher, or most other appliances. Five years of trouble-free service are about the best you can expect—then repairs are likely to be needed. There are some steps you can take to prolong life. One is to make sure that the machine is properly balanced. There should be leveling lugs at all four corners, and if they don't quite do the job, use wood shims. Even then, the spinning action may knock the machine off balance, so leveling every year or so is a wise investment of time. Whenever the washer starts to thump or travel, it is time to check the level, or check the balance of the load being spun.

Unbalanced loads, overloading, using too much or the wrong type of detergent, and general disregard of the manufacturer's instructions are shortcuts to the appliance graveyard. When a problem occurs, again, always check the obvious. A special problem for washers is a small item such as a sock or a handkerchief that somehow gets tossed out of the washer basket and lodges in the mechanism. To look for such objects, you need to remove the top or back panel of the machine. Check the manufacturer's manual, or visually inspect to see how this is done. If a piece of clothing is caught beneath the agitator, remove the agitator by lifting it out or by turning the top section clockwise and lifting it out. Then remove the obstruction.

Diagnostic Chart—Dry and Steam Irons

Symptoms	*Cause*	*Treatment*
No heat	No power at outlet	Check outlet, fuse; repair or replace if necessary
	Defective cord, plug	Repair or replace
	Broken lead in iron	Repair or replace
	Loose connection	Clean and tighten
	Loose thermostat control knob	Replace knob and tighten
	Defective thermostat	Replace
	Defective heating element	Replace element if separate; replace bottomplate if attached
Insufficient heat	Low line voltage*	Check voltage at outlet; notify utility if deficient
	Incorrect thermostat setting	Adjust thermostat
	Defective thermostat	Replace
	Loose connection	Clean and tighten
Excessive heat	Incorrect thermostat setting	Adjust thermostat
	Defective thermostat	Replace
Blisters on bottomplate	Excessive heat	Correct condition (above): repair or replace
Water leakage	Defective seam or tank weld	Replace tank
	Inadequate tank seal	Reseal with proper sealer
	Damaged gasket	Replace gasket
No steam	Thermostat set too low	Set control higher
	Valve in off position	Turn to correct position
	Dirty or plugged valves or holes	Clean
"Spitting"	Incorrect thermostat setting	Set thermostat higher
	Excessive mineral deposits	Clean
	Overfilling	Drain, be more careful
Bad spray (spray irons)	Defective plunger	Replace
Stains on clothes	Starch on bottomplate	Rub bottomplate with damp cloth, polish with a dry cloth
	Foreign matter in water	Use distilled water
	Sediment in tank	Clean with vinegar

Diagnostic Chart—Dry and Steam Irons (continued)

Symptoms	*Cause*	*Treatment*
Tears clothes	Rough spot, nick, scratch, or burr on bottomplate	Remove with fine emery cloth, then buff or polish
Sticks to clothes	Dirty bottomplate	Clean
	Excessive starch in clothes	Iron at a lower temperature; use less starch

*As noted previously; Low and high line voltages spell trouble for many devices, and the power company is required to correct such situations.

Diagnostic Chart—Toasters

Symptoms	*Cause*	*Treatment*
No heat	No power at outlet	Check outlet, fuse; repair or replace if necessary
	Defective cord	Repair or replace
	Loose connection	Clean, tighten
	Switch not making contact	Repair or replace
	Elements burned out	Replace
Toast will not stay down	Hold-down latch not locking	If bent, straighten; if binding, clear to allow free operation
	Bind in toast carriage	Clear cause of bind
	Broken latch spring	Replace
Toast will not pop up	Bind in toast carriage	Clear cause of bind
	Release latch binds	Clear cause of bind
	Broken spring	Replace
Uneven toasting	Broken or bent element	Replace element

Diagnostic Chart—Coffee Makers

Symptoms	*Cause*	*Treatment*
Doesn't operate	No voltage at outlet	Check outlet, fuse; repair or replace if necessary
	Defective cord	Repair or replace
	Defective pump heater	Replace
Gets warm but doesn't percolate	Defective pump heater	Replace
	Defective thermostat	Replace
	Incorrect setting of thermostat	Reset
Slow in brewing	Low line voltage	Check voltage: if low, notify power company
Coffee tastes bitter	Accumulated residue inside pot	Clean with baking soda solution or other cleaner
Weak coffee	Control incorrectly set	Reset
	Using hot water to start	Use cold water
	Pump valve stuck	Clean valve or replace if damaged
Coffee boils	Incorrect thermostat setting	Reset
	Defective thermostat	Replace

Diagnostic Chart—Electric Frying Pans

Symptoms	*Cause*	*Treatment*
No heat	No power at outset	Check outlet, fuse; repair or replace if necessary
	Defective cord	Repair or replace
	Broken heating element	Replace
	Defective thermostat	Replace
	Poor connection	Clean and tighten
No heat control	Defective thermostat	Replace
Shocks user	Grounded unit	Replace defective part
	Wire touching frame	Locate and reinsulate or replace wire
Food sticks	Excessive cooking temperature	Lower thermostat setting
	Pan is not seasoned	Season pan; heat for half hour with shortening

Diagnostic Chart—Waffle Irons

Symptoms	*Cause*	*Treatment*
No heat	No power at outlet	Check outlet, fuse; repair or replace if necessary
	Defective cord	Repair or replace
	Damaged heating element	Replace
	Broken hinge wire	Replace
	Defective thermostat	Replace
Blows fuses	Shorted cord	Replace
	Shorted heater element	Replace
	Shorted wiring	Reinsulate
Too hot	Check thermostat setting (maximum temperature 520° F.)	Reset thermostat, replace if faulty
Waffles stick	Improperly seasoned grid	Operate for half-hour with cooking oil—no batter—on grids
	Insufficient shortening in batter	Add more cooking oil
	Too much sugar in batter	Correct error
	Opening griddle too soon	Be patient!

Diagnostic Chart—Ranges and Ovens

Symptoms	*Cause*	*Treatment*
No heat at surface element	No power	Check outlet, fuses; repair or replace if necessary
	Thermostat control	Clean or replace
	Heating element	Check connection; replace if necessary
	Internal wiring	Check connections
	Control switch	Replace

Diagnostic Chart—Ranges and Ovens (continued)

Symptoms	*Cause*	*Treatment*
Partial heat at element	Low line voltage	Add house power (check with your utility)
	Thermostat control	Clean or replace
	Heating element	Check connection; replace if necessary
	Internal wiring	Check connections
	Control switch	Replace
No heat at broiler	Low line voltage	Add house power
	Thermostat control	Clean or replace
	Heating element in broiler	Check connections; replace if necessary
	Internal wiring	Check connections
	Control switch	Replace
No heat at oven	Low line voltage	Add house power
	Oven heating element	Check connections; replace if necessary
	Internal wiring	Check connections
	Control switch	Replace

Diagnostic Chart—Portable Heaters

Symptoms	*Cause*	*Treatment*
No heat	No power to unit	Check fuse, outlet; repair or replace if necessary
	Defective cord, plug	Repair or replace
	Defective switch	Replace
	Defective thermostat	Replace
	Defective heating element	Replace
Low heat	Incorrect thermostat setting	Adjust thermostat
	Defective heating element	Replace
Won't shut off	Defective switch	Replace
	Defective thermostat	Replace
	Shorted wiring	Locate, separate wires, reinsulate
Fan doesn't run	Loose connection	Locate and tighten
	Jammed fan blade	Straighten
	Frozen motor bearing	Free armature, lubricate bearing
	Burned-out motor	Replace
Excessively noisy	Fan blade hitting obstruction	Straighten blade or clear obstruction away from blade
	Worn motor bearings	Replace
Shocks user	Defective wiring	Clear bare wire from frame and insulate; replace wire
	Heating element touching reflector	Repair or replace element; bend back dented reflector

Diagnostic Chart—Rotisseries

Symptoms	*Cause*	*Treatment*
Won't operate	No power at outlet	Check outlet, fuse; repair or replace if necessary
	Defective cord	Repair or replace
	Defective switch	Replace
	Defective timer switch	Replace
Heats, but motor doesn't run	Defective motor	Replace
	Loose connections	Clean and tighten
	Stuck gearing	Clean gears
	Defective motor switch	Replace
Motor runs, but oven doesn't heat	Defective heating element	Replace
	Defective heater switch	Replace
Incorrect heat	Defective thermostat	Replace

Diagnostic Chart—Dryers

Symptoms	*Cause*	*Treatment*
Won't operate	No power	Check outlet, fuse; repair or replace if necessary
	Timer contacts	Clean or replace
	Motor	Repair or replace
	Door switch	Remove obstruction or replace
Runs but no heat	Timer contacts	Clean or replace
	Thermostat	Clean or replace
	Heating element	Check connections; replace if necessary
Heats but won't turn	Drive belt	Tighten or replace
	Transmission	Replace defective gear
	Bad motor or loose connection	Clean and tighten
Overheats	Drive belt	Tighten or replace
	Thermostat	Clean or replace
Timer doesn't move	Timer contacts	Clean or replace
	Door switch	Remove obstruction or replace

Diagnostic Chart—Dishwashers

Symptoms	*Cause*	*Treatment*
Doesn't operate	No power	Check outlet, fuse; repair or replace as necessary
	Timer contacts	Clean or replace
	Defective motor	Replace brushes or motor
	Door switch	Remove obstruction or replace
No water	Timer contacts	Clean or replace
	Defective solenoid	Replace
	Supply valve off	Open up; unclog
	Door switch	Remove obstruction or replace

Diagnostic Chart—Dishwashers (continued)

Symptoms	*Cause*	*Treatment*
Won't drain	Timer contacts	Clean or replace
	Defective solenoid	Replace
	Drive belt	Tighten or replace
	Level control	Clean or replace
	Pump	Remove obstruction or replace
Won't feed detergent	Timer contacts	Clean or replace
	Defective solenoid	Replace
Won't dry	Timer contacts	Clean or replace
	Thermostat	Clean or replace
	Heating element	Replace

Diagnostic Chart—Mixers

Symptoms	*Cause*	*Treatment*
Doesn't work at all	No power at outlet	Check fuse, outlet; repair or replace if necessary
	Defective cord	Repair or replace
	Worn brushes	Replace
	Broken field cord	Replace
	Broken armature winding	Replace armature
	Defective switch	Replace
Doesn't run, blows fuses	Bent shaft jamming armature	Straighten or replace shaft
	Defective armature or field coil	Replace
	Shorted cord	Repair or replace
Motor runs hot	Bind in shaft	Clear bind
	Shorted winding in armature	Replace armature
	Shorted field coil	Replace
Motor runs, beaters don't turn	Bent, damaged beaters	Replace
	Stripped gears	Replace
Erratic operation	Worn brushes	Replace
	Loose connection	Clean and tighten
	Defective switch	Replace
Slow speed, weak power	Incorrect setting	Adjust speed control
	Worn brushes	Replace
	Bind in shaft	Clear bind
Too noisy	Armature hitting field	Replace worn bearings
	Bent cooling fan blade	Straighten
	Dry gears or bearing	Lubricate

Diagnostic Chart—Blenders

Symptoms	*Cause*	*Treatment*
Motor won't run	Problem at outlet	Check fuse, outlet, wiring; repair or replace as necessary
	Loose connections	Clean and tighten

Diagnostic Chart—Blenders (continued)

Symptoms	*Cause*	*Treatment*
Motor won't run	Defective cord	Repair or replace
	Defective switch	Repair or replace
	Burned-out motor	Replace armature or field coil
	Frozen bearings	Free, lubricate
	Armature hitting field	Replace worn bearings
Motor runs, blade doesn't turn	Broken belt (on some models)	Replace belt
	Incorrect placement of container	Relocate on base
	Defective motor coupling	Replace
Runs at high speed only	Defective switch	Replace
	Open resistor	Replace
	Defective field coil	Replace
Blade damaged	Hitting ice cubes, spoon, bones, etc.	Replace blade
Container leaks	Cracked glass jar	Replace
	Poor seal	Tighten bushing or replace gasket

Diagnostic Chart—Vacuum Cleaners

Symptoms	*Cause*	*Treatment*
Motor doesn't run	No power	Check outlet, fuse, wiring; repair or replace if necessary
	Loose connection	Clean and tighten
	Defect in cord, plug	Repair or replace
	Defective switch	Replace
	Worn brushes	Replace
	Jammed fan	Free; if bent or damaged, replace
	Frozen bearings	Clean and lubricate; if worn, replace
Motor starts and stops	Intermittent break in cord	Locate and repair or replace
	Loose connection in cleaner	Check all connections; repair
	Defective switch	Replace
	Loose connection in motor	Check motor; tighten connection
Motor runs too slow, no power	Foreign object caught on fan or armature	Remove object
	Misaligned, tight motor bearings	Realign, retighten
	Poor brush contact	Correct contact or replace brushes
Motor runs too fast	Overfilled dust bag	Replace
	Fan loose on shaft, not turning	Check its balance, tighten
Motor sparks	Dirty commutator (oil or dirt)	Clean with fine sandpaper
	Worn brushes	Replace
	Incorrect brush seating	Correct seating
Motor too noisy	Foreign matter in motor	Clean out
	Fan damaged	Replace

Diagnostic Chart—Vacuum Cleaners (continued)

Symptoms	*Cause*	*Treatment*
Poor pickup	Worn or damaged attachments	Check attachments for leakage and replace as necessary
	Incorrect nozzle adjustment for carpet nap	Adjust for correct contact
	Leaky hose	Check for air leaks; repair or replace
	Clogged hose	Blow or push out obstruction
	Overfilled dust bag	Replace
	Clogged exhaust port	Clear
Dust leakage	Holes in dust bag	Replace
	Incorrectly installed dust bag	Check owner's manual
	Old, dirty dust bag	Replace
	Defective sealing	Replace seal

Diagnostic Chart—Polisher-Scrubbers

Symptoms	*Cause*	*Treatment*
Will not run	No power at outlet	Check outlet, fuse, wiring; repair or replace if necessary
	Loose connection	Clean and tighten
	Defective cord	Check for break in cord, plug; repair or replace
	Loose connection in plug at handle	Tighten connection
	Handle switch defective	Replace
	No voltage; low voltage	Check voltage at motor terminals; notify utility if inadequate
	Motor failure	Repair or replace
Motor runs, polishing brushes turn slowly or not at all	Worn or loose belt	Replace
Motor runs hot	Bearing problem	Check alignment of bearing; lubricate
	Poor ventilation	Check vent openings; clean out
Motor smokes	Motor failure	Replace
Excessive noise	Dirt in air gap	Clean out
	Worn bearings	Replace
	Unbalanced rotor	Balance or replace
	Excessive end play	Check armature clearance
	Worn gears or spindle bearings	Replace
	Improper lubrication	Lubricate
Excessive vibration	One worn polishing brush	Replace both polishing brushes
Floor spotty and streaked after scrubbing	Dirty suds	Remove dirty suds with clean rag
	Dirty brushes	Clean by washing with soap and warm water; shake out and let dry before re-using

Diagnostic Chart—Polisher-Scrubbers (continued)

Symptoms	*Cause*	*Treatment*
Floor sticky, slippery or shows footprints after polishing	Inferior grade of wax	Use a better grade
	Too much wax	Use lighter applications
	Wax not rubbed in well or allowed to dry properly	Practice (and patience) makes perfect
Ruts or grooves cut in floor when sanding	Sandpaper too coarse	Use a finer grade
	Improper use of sanding attachment; sanding in wrong direction	Sand diagonally or across grain of the floor

Diagnostic Chart—Refrigerators

Symptoms	*Cause*	*Treatment*
Won't operate	No power	Check outlet, fuse; repair or replace if necessary
	Low line voltage	Add house power
	Thermostat control	Clean or replace
	Compressor, motor	Repair or replace
	Internal wiring	Check connections
Too warm inside	Thermostat control	Clean or replace if necessary
	Compressor, motor	Repair or replace
	Fan motor	Repair or replace
	Defrost thermostat	Clean or replace
Won't freeze	Low line voltage	Add house power
	Thermostat control	Clean or replace
	Compressor, motor	Repair or replace
	Fan motor	Repair or replace
	Internal wiring	Check connections
Runs continuously	Low line voltage	Add house power
	Thermostat control	Clean or replace
	Compressor, motor	Repair or replace
	Fan motor	Repair or replace
Excess frost	Thermostat control	Clean or replace
	Compressor, motor	Repair or replace
	Freezer control	Repair or replace
Fresh foods freeze	Thermostat control	Clean or replace
	Compressor	Repair or replace
Erratic cooling	Leaking coolant	Recharge

Diagnostic Chart—Clothes Washers

Symptoms	*Cause*	*Treatment*
Washer doesn't fill	No power	Check outlet, fuse; repair or replace as necessary
	Timer contacts	Clean or replace
	Defective solenoid	Replace
	Supply valve off	Open up; unclog
	Level control	Clean or replace

Diagnostic Chart—Clothes Washers (continued)

Symptoms	*Cause*	*Treatment*
Motor won't run	No power	Check outlet, fuse; repair or replace if necessary
	Loose connections	Clean and tighten
	Timer contacts	Clean or replace
	Motor	Repair or replace
	Door switch	Remove obstruction or replace
Agitator doesn't work	Timer contacts	Clean or replace
	Motor	Repair or replace
	Defective solenoid	Replace
	Drive belt	Tighten or replace
	Transmission	Replace defective part
	Door switch	Remove obstruction or replace
Doesn't spin	Timer contacts	Clean or replace
	Motor	Repair or replace
	Defective solenoid	Replace
	Drive belt	Tighten or replace
	Transmission	Replace defective part
	Level control	Clean or replace
	Door switch	Remove obstruction or replace
Doesn't rinse	Timer contacts	Clean or replace
	Defective solenoid	Replace
	Supply valve off	Open up; unclog
	Level control	Clean or replace
	Door switch	Remove obstruction or replace
Doesn't drain	Timer contacts	Clean or replace
	Drive belt	Tighten or replace
	Level control	Clean or replace
	Door switch	Remove obstruction or replace
	Pump	Remove obstruction or replace
Timer doesn't move	Timer contacts	Clean or replace
	Door switch	Remove obstruction or replace

Index

Page numbers in **bold** type indicate information found in illustrations.
Page numbers followed by "t" indicate information found in tables.